Menu Math

THE HAMBURGER HUT (+,-)

REM 102A

WRITTEN BY: Kitty Scharf
Barbara Johnson

A TEACHING RESOURCE FROM

To find Remedia products in a store near you, visit:
www.rempub.com/stores

REMEDIA PUBLICATIONS, INC.
15887 NORTH 76TH STREET • SUITE 120 • SCOTTSDALE, AZ • 85260

This product utilizes innovative strategies and proven methods to improve student learning. The product is based upon reliable research and effective practices that have been replicated in classrooms across the United States. Information regarding the research basis is provided on our website at www.rempub.com/research

MENU MATH

<u>TO THE TEACHER</u>

MENU MATH books are designed to build basic math skills through the use of a real-life, contemporary situation—looking at a menu and figuring costs involved when eating in a restaurant. By answering questions which have realistic circumstances, students are often able to better understand the practical reasons for learning math.

This approach has proven successful in helping motivate the sometimes reluctant learner. Most students show a natural interest in materials featuring topics familiar to them as a part of their everyday lives.

Skill pages in this book include clearly stated objectives and are sequential. Post-tests appear at the end of each section. In most cases, examples are provided for the students. However, so that a page may be used as a post-test, examples have been left off some pages.

<u>SUGGESTIONS</u>

- Duplicate pages.
- Set up MENU MATH Learning Centers.
- Have students create their own story problems and checks.
- Write additional story problems and checks using your students' names.
- Make a booklet of selected skill pages for individual students. Have the student design a cover for the booklet—draw a hamburger, cola, etc.
- Cut out the restaurant checks. They may be used along with play money to help students learn how to make change.
- Allow students to use calculators to check answers.
- Purchase additional copies of the menus.

KNOWLEDGE OF MATH AND MONEY IS IMPORTANT TO SUCCESS AND INDEPENDENCE. WE HOPE YOUR STUDENTS WILL ENJOY AND LEARN FROM THE REAL-LIFE EXPERIENCES PROVIDED IN THIS BOOK.

Name ______________________________

Using The Menu

Use the Hamburger Hut Menu to answer the questions.

1. How much does a Jumbo Burger cost? __________
2. How much for an order of toast? __________
3. Meat loaf costs how much? __________
4. How much does a chicken sandwich cost? __________
5. What kind of potatoes come with liver & onions? __________
6. How much for coconut cake? __________
7. What comes with ham & eggs? ________________________
8. What kind of cheese is on the Bacon Burger? __________
9. How much is a Waldorf Salad? __________
10. What kind of potatoes come with the Fish Dinner? __________
11. How much for Onion Soup? __________
12. What kinds of juice are available at the Hamburger Hut? ________________________

____

____

Name ______________________________

Introducing decimal notation for money

Example:

38¢ can be written as...

$.38 ← 8 pennies
- 3 dimes
- decimal point
- dollar sign

Write the amounts below using the dollar sign ($) and decimal point (.).

1. Six cents ____________
2. Eight cents ____________
3. Ten cents ____________
4. Fifteen cents ____________
5. Forty-three cents ____________
6. One dollar and twenty-five cents ____________
7. Four dollars and thirteen cents ____________
8. Fourteen dollars and five cents ____________
9. Sixty-three dollars and twenty cents ____________
10. Eighty-four dollars and six cents ____________

Name ______________________________

Adding two 2-digit numbers, no regrouping.

Example:

$$\begin{array}{r} \$\ .53 \\ +\ \underline{.12} \\ \$\ .65 \end{array}$$

1.	$\begin{array}{r} \$\ .12 \\ +\ \underline{.04} \end{array}$	$\begin{array}{r} \$\ .17 \\ +\ \underline{.02} \end{array}$	$\begin{array}{r} \$\ .35 \\ +\ \underline{.04} \end{array}$	$\begin{array}{r} \$\ .12 \\ +\ \underline{.10} \end{array}$
2.	$\begin{array}{r} \$\ .18 \\ +\ \underline{.40} \end{array}$	$\begin{array}{r} \$\ .11 \\ +\ \underline{.28} \end{array}$	$\begin{array}{r} \$\ .43 \\ +\ \underline{.23} \end{array}$	$\begin{array}{r} \$\ .36 \\ +\ \underline{.12} \end{array}$
3.	$\begin{array}{r} \$\ .10 \\ +\ \underline{.28} \end{array}$	$\begin{array}{r} \$\ .12 \\ +\ \underline{.27} \end{array}$	$\begin{array}{r} \$\ .48 \\ +\ \underline{.11} \end{array}$	$\begin{array}{r} \$\ .62 \\ +\ \underline{.23} \end{array}$
4.	$\begin{array}{r} \$\ .43 \\ +\ \underline{.22} \end{array}$	$\begin{array}{r} \$\ .26 \\ +\ \underline{.21} \end{array}$	$\begin{array}{r} \$\ .52 \\ +\ \underline{.25} \end{array}$	$\begin{array}{r} \$\ .48 \\ +\ \underline{.21} \end{array}$
5.	$\begin{array}{r} \$\ .18 \\ +\ \underline{.31} \end{array}$	$\begin{array}{r} \$\ .44 \\ +\ \underline{.24} \end{array}$	$\begin{array}{r} \$\ .72 \\ +\ \underline{.16} \end{array}$	$\begin{array}{r} \$\ .47 \\ +\ \underline{.32} \end{array}$
6.	$\begin{array}{r} \$\ .20 \\ +\ \underline{.36} \end{array}$	$\begin{array}{r} \$\ .63 \\ +\ \underline{.23} \end{array}$	$\begin{array}{r} \$\ .74 \\ +\ \underline{.25} \end{array}$	$\begin{array}{r} \$\ .67 \\ +\ \underline{.22} \end{array}$

Name ______________________________

Adding two 2-digit numbers, no regrouping.

1.	$ .15 + .02	$ .16 + .03	$ .21 + .05	$ .13 + .15
2.	$ .23 + .42	$ .15 + .23	$ .46 + .12	$ .41 + .36
3.	$ .20 + .68	$ .32 + .14	$ .41 + .26	$ .63 + .25
4.	$ .25 + .23	$ .28 + .51	$ .56 + .22	$ .43 + .26
5.	$ .33 + .15	$ .82 + .16	$ .74 + .22	$ .65 + .24
6.	$ .92 + .05	$ .84 + .14	$ .22 + .46	$ .36 + .32

Name ____________________________________

Adding three 2-digit numbers, no regrouping.

Example:

$$\begin{array}{r} \$\ .12 \\ .15 \\ +\ \underline{.10} \\ \$\ .37 \end{array}$$

1.	\$ .22 .43 + .11	\$ .24 .62 + .13	\$ .21 .18 + .30	\$ .31 .32 + .04
2.	\$ .11 .23 + .42	\$ .23 .13 + .63	\$ .52 .14 + .11	\$ .44 .10 + .20
3.	\$ .24 .41 + .14	\$ .25 .31 + .42	\$ .33 .32 + .21	\$ .46 .10 + .11
4.	\$ .32 .15 + .12	\$ .17 .10 + .21	\$ .56 .12 + .31	\$ .81 .12 + .06
5.	\$ .17 .22 + .40	\$ .25 .53 + .11	\$ .62 .21 + .03	\$ .54 .13 + .22
6.	\$ .16 .41 + .40	\$ .21 .52 + .26	\$ .43 .15 + .11	\$ .55 .04 + .30

Name ______________________________

Adding three 2 digit numbers, no regrouping.

1.	$ .14 .21 + .14	$ .15 .40 + .43	$ .36 .21 + .41	$ .23 .36 + .30
2.	$ .25 .40 + .13	$ .27 .12 + .20	$ .33 .32 + .12	$ .43 .12 + .24
3.	$ .13 .24 + .22	$ .33 .34 + .12	$ .53 .22 + .14	$ .52 .20 + .13
4.	$ .25 .23 + .40	$ .16 .23 + .40	$ .61 .32 + .05	$ .16 .20 + .43
5.	$ .18 .30 + .51	$ .23 .14 + .42	$ .37 .20 + .32	$ .56 .22 + .21
6.	$ .32 .41 + .25	$ .22 .61 + .14	$ .63 .21 + .15	$ .80 .15 + .04

Name ______________________________

Adding two 3-digit numbers, no regrouping.

Example:
$ 4.23
+ 2.34
$ 6.57

1.	$ 2.32 + 1.13	$ 3.43 + 1.24	$ 2.16 + 4.03	$ 4.32 + 1.22
2.	$ 3.18 + 2.01	$ 2.26 + 2.13	$ 4.03 + 4.14	$ 3.21 + 2.07
3.	$ 6.32 + 3.22	$ 2.15 + 4.13	$ 4.07 + 3.01	$ 5.03 + 4.21
4.	$ 4.64 + 3.21	$ 6.25 + 3.24	$ 5.52 + 2.26	$ 4.53 + 3.06
5.	$ 7.21 + 2.13	$ 4.06 + 4.00	$ 7.15 + 1.62	$ 4.80 + 1.15
6.	$ 6.12 + 3.06	$ 6.52 + 1.35	$ 8.02 + 1.54	$ 7.24 + 2.15

Name ______________________________

Adding two 3-digit numbers, no regrouping.

1.	$ 2.64 + 3.15	$ 4.25 + 4.43	$ 5.82 + 3.16	$ 3.36 + 5.53
2.	$ 6.25 + 1.72	$ 8.24 + 1.44	$ 6.33 + 3.44	$ 2.82 + 7.16
3.	$ 3.82 + 4.16	$ 5.36 + 2.63	$ 2.81 + 6.17	$ 4.22 + 4.56
4.	$ 7.21 + 2.78	$ 6.37 + 2.41	$ 2.48 + 7.51	$ 6.62 + 3.24
5.	$ 6.13 + 2.74	$ 5.46 + 2.43	$ 4.86 + 3.12	$ 7.47 + 2.52
6.	$ 6.21 + 2.46	$ 8.52 + 1.36	$ 7.81 + 2.18	$ 8.43 + 1.25

Name ______________________________

Adding three 3-digit numbers, no regrouping.

Example:

```
  $ 3.14
    4.12
  + 2.11
  $ 9.37
```

```
1.  $ 3.40     $ 2.12     $ 3.15     $ 2.10
      2.32       3.61       4.10       1.53
    + 4.12     + 2.24     + 2.12     + 6.12

2.  $ 2.05     $ 1.42     $ 1.46     $ 1.22
      1.23       2.32       3.11       3.22
    + 4.10     + 5.11     + 4.21     + 4.13

3.  $ 1.36     $ 3.01     $ 2.51     $ 4.12
      2.23       2.06       1.22       1.15
    + 1.00     + 3.51     + 3.13     + 3.11

4.  $ 2.00     $ 4.72     $ 6.12     $ 3.41
      3.35       1.25       1.25       2.16
    + 3.53     + 3.01     + 1.20     + 1.22

5.  $ 1.61     $ 5.20     $ 4.12     $ 5.21
      3.02       2.13       1.04       1.06
    + 3.04     + 2.12     + 2.10     + 3.02

6.  $ 4.02     $ 3.36     $ 5.00     $ 6.00
      2.34       2.21       1.25       2.05
    + 2.43     + 2.40     + 2.62     + 1.13
```

Name ____________________

Adding three 3-digit numbers, no regrouping.

1.	$ 2.31 1.42 + 3.20	$ 2.23 1.41 + 2.05	$ 3.00 2.10 + 1.46	$ 4.84 1.03 + 2.12
2.	$ 4.14 1.21 + 1.04	$ 2.15 3.51 + 2.23	$ 1.24 2.11 + 4.32	$ 6.11 1.21 + 2.32
3.	$ 6.12 1.33 + 1.04	$ 1.63 2.30 + 3.01	$ 3.12 2.02 + 1.43	$ 4.41 2.01 + 2.33
4.	$ 2.03 1.11 + 2.40	$ 4.01 1.23 + 3.72	$ 6.11 1.42 + 2.32	$ 3.15 1.21 + 1.01
5.	$ 1.22 3.33 + 1.04	$ 2.02 3.50 + 3.22	$ 2.06 2.01 + 4.11	$ 1.06 1.71 + 2.22
6.	$ 1.52 2.11 + 3.06	$ 4.23 2.12 + 2.03	$ 5.12 1.42 + 2.03	$ 1.06 1.01 + 4.12

Name ________________________________

Adding two 2-digit numbers, one-step regrouping.

Example:

```
     1
  $ .28
  + .14
  $ .42
```

1.	$ .27 + .13	$ .44 + .16	$ .53 + .28	$ .19 + .44
2.	$ .29 + .31	$ .24 + .17	$ .36 + .16	$ .59 + .28
3.	$ .26 + .16	$ .27 + .14	$ .59 + .24	$ .38 + .14
4.	$ .27 + .29	$ .72 + .18	$ .15 + .65	$ .36 + .25
5.	$ .29 + .16	$ .28 + .36	$ .64 + .18	$ .27 + .23
6.	$ .72 + .19	$ .64 + .26	$ .59 + .13	$ .52 + .39

Name ______________________________

Adding two 2-digit numbers, one-step regrouping.

1.	\$ .43 + .18	\$ .22 + .19	\$ .34 + .16	\$ .28 + .24
2.	\$ .62 + .18	\$ .25 + .25	\$ .36 + .28	\$ .77 + .15
3.	\$ .27 + .24	\$ .33 + .19	\$ .76 + .16	\$ .29 + .34
4.	\$ .23 + .28	\$ .17 + .25	\$ .43 + .48	\$ .36 + .48
5.	\$ .52 + .18	\$ .64 + .17	\$ .54 + .38	\$ .76 + .17
6.	\$ .55 + .39	\$ .27 + .18	\$ .56 + .15	\$ .39 + .25

Name ______________________________

Adding two 2-digit numbers, two-step regrouping.

Example:

```
   1
$ .57
+ .64
$1.21
```

1.	$ .47 + .83	$ .26 + .94	$ .65 + .65	$ .27 + .84
2.	$ .95 + .26	$ .88 + .84	$ .78 + .94	$ .88 + .34
3.	$ .64 + .76	$ .56 + .94	$ .87 + .43	$ .75 + .57
4.	$ .57 + .89	$ .98 + .76	$ .57 + .94	$ .88 + .88
5.	$ .76 + .75	$ .95 + .28	$ .73 + .89	$ .55 + .68
6.	$ .56 + .56	$ .67 + .35	$ .87 + .46	$ .94 + .98

Name ______________________________

Adding two 2-digit numbers, two-step regrouping.

1.	\$ .75 + .58	\$ .86 + .76	\$ .59 + .89	\$ .23 + .99
2.	\$ .56 + .88	\$ .85 + .89	\$ .65 + .78	\$ .99 + .99
3.	\$ .52 + .59	\$ .84 + .78	\$ .48 + .98	\$ .45 + .76
4.	\$ .18 + .86	\$ .84 + .59	\$ .39 + .78	\$ .38 + .98
5.	\$ .79 + .78	\$ .84 + .57	\$ .66 + .86	\$ .96 + .45
6.	\$ .49 + .79	\$ .74 + .69	\$ .37 + .83	\$ .86 + .27

Name ________________________________

Adding two 3-digit numbers, one-step regrouping.

Example:

```
  1
$ 6.38
+ 2.27
$ 8.65
```

1.

```
$ 6.25      $ 1.35      $ 2.46      $ 7.87
+ 2.36      + 1.38      + 3.27      + 1.03
```

2.

```
$ 5.24      $ 2.67      $ 7.46      $ 1.23
+ 2.38      + 1.29      + 1.26      + 1.37
```

3.

```
$ 4.52      $ 6.36      $ 3.77      $ 1.58
+ 2.38      + 1.48      + 2.07      + 2.18
```

4.

```
$ 2.66      $ 1.15      $ 2.17      $ 5.64
+ 3.24      + 1.35      + 2.49      + 2.28
```

5.

```
$ 3.18      $ 6.17      $ 1.45      $ 4.18
+ 2.17      + 2.69      + 3.36      + 2.39
```

6.

```
$ 7.15      $ 8.24      $ 7.49      $ 5.58
+ 2.06      + 1.36      + 2.09      + 3.27
```

Name ______________________________

Adding two 3-digit numbers, one-step regrouping.

1.	$ 5.27 + 2.64	$ 8.32 + 1.28	$ 1.59 + 2.27	$ 2.39 + 1.29
2.	$ 8.27 + 1.58	$ 1.64 + 1.09	$ 2.77 + 1.18	$ 4.25 + 3.65
3.	$ 5.23 + 1.68	$ 1.16 + 2.57	$ 4.31 + 4.09	$ 2.57 + 6.33
4.	$ 6.38 + 2.44	$ 3.26 + 2.65	$ 7.04 + 2.57	$ 6.48 + 1.38
5.	$ 2.63 + 4.17	$ 3.26 + 4.64	$ 2.07 + 5.46	$ 5.63 + 2.08
6.	$ 5.26 + 3.49	$ 7.86 + 1.06	$ 6.59 + 1.11	$ 4.24 + 2.27

Name ______________________________

Adding two 3-digit numbers, two-step regrouping.

Example:
$ 5.36
+ 2.77
$ 8.13

---

1. $ 6.36 + 2.87 | $ 2.69 + 5.83 | $ 7.24 + 1.88 | $ 4.58 + 3.94

2. $ 7.95 + 1.47 | $ 2.13 + 1.87 | $ 4.92 + 1.89 | $ 7.83 + 1.98

3. $ 5.24 + 2.97 | $ 3.58 + 5.42 | $ 2.37 + 4.69 | $ 6.52 + 2.98

4. $ 4.73 + 2.59 | $ 4.47 + 3.87 | $ 1.07 + 2.93 | $ 3.38 + 4.67

5. $ 2.58 + 3.67 | $ 2.98 + 2.02 | $ 4.87 + 2.85 | $ 5.97 + 1.58

6. $ 5.53 + 3.68 | $ 7.21 + 1.79 | $ 4.57 + 2.53 | $ 5.49 + 3.99

Name ________________________________

Adding two 3-digit numbers, two-step regrouping.

1.	\$ 7.67 + 1.96	\$ 3.16 + 2.87	\$ 3.65 + 3.35	\$ 3.68 + 5.67
2.	\$ 4.75 + 2.35	\$ 2.88 + 3.88	\$ 3.65 + 1.57	\$ 1.87 + 1.56
3.	\$ 4.52 + 2.69	\$ 6.55 + 2.55	\$ 5.38 + 1.68	\$ 2.87 + 5.37
4.	\$ 3.55 + 3.49	\$ 5.68 + 3.54	\$ 2.69 + 1.89	\$ 2.32 + 2.68
5.	\$ 5.99 + 2.01	\$ 2.53 + 6.47	\$ 4.18 + 4.82	\$ 2.98 + 1.35
6.	\$ 5.89 + 2.93	\$ 1.53 + 2.77	\$ 4.93 + 4.89	\$ 3.83 + 2.99

Name ______________________________

Adding two 3-digit numbers, two or more regrouping steps.

Example:

```
    1 1
$   9.54
+   3.77
$  13.31
```

1.	\$ 4.36 + 5.76	\$ 5.58 + 8.94	\$ 6.22 + 9.78	\$ 7.96 + 6.79
2.	\$ 5.49 + 6.76	\$ 2.98 + 8.47	\$ 5.44 + 9.89	\$ 6.28 + 7.94
3.	\$ 7.98 + 5.49	\$ 3.74 + 8.67	\$ 6.47 + 9.53	\$ 8.38 + 7.93
4.	\$ 3.68 + 9.48	\$ 9.52 + 3.78	\$ 6.48 + 5.64	\$ 6.58 + 6.78
5.	\$ 2.98 + 9.24	\$ 3.65 + 7.36	\$ 5.44 + 8.66	\$ 8.59 + 7.64
6.	\$ 5.96 + 9.78	\$ 7.49 + 3.88	\$ 6.69 + 7.81	\$ 9.49 + 3.84

Name ________________________________

Adding two 3-digit numbers, two or more regrouping steps.

1.	\$ 6.58 + 3.84	\$ 8.79 + 6.85	\$ 6.15 + 9.98	\$ 7.28 + 4.96
2.	\$ 6.29 + 4.89	\$ 5.86 + 5.47	\$ 9.99 + 2.26	\$ 7.87 + 3.56
3.	\$ 9.72 + 9.69	\$ 7.48 + 8.56	\$ 7.85 + 2.95	\$ 3.84 + 8.37
4.	\$ 3.89 + 8.24	\$ 8.64 + 5.86	\$ 6.48 + 8.88	\$ 7.56 + 6.75
5.	\$ 2.07 + 8.97	\$ 5.55 + 8.78	\$ 7.28 + 2.99	\$ 6.27 + 4.93
6.	\$ 8.66 + 8.69	\$ 7.99 + 2.22	\$ 6.57 + 6.58	\$ 2.11 + 9.99

Name ______________________________

Adding two 4-digit numbers, two or more regrouping steps.

Example:

```
  1 1 1
$ 62.35
+ 29.85
$ 92.20
```

1.

```
$ 23.65    $ 30.56    $ 47.55    $ 32.57
+ 14.66    + 28.74    + 38.45    + 18.46
```

2.

```
$ 30.46    $ 20.28    $ 33.44    $ 24.68
+ 23.74    + 14.72    + 16.78    + 55.74
```

3.

```
$ 60.82    $ 73.56    $ 18.68    $ 63.44
+ 29.29    + 11.74    + 12.56    + 27.97
```

4.

```
$ 63.25    $ 21.07    $ 45.63    $ 22.75
+ 26.76    + 18.98    + 27.67    + 14.39
```

5.

```
$ 68.76    $ 27.48    $ 25.79    $ 57.73
+ 11.69    + 13.94    + 18.81    + 17.77
```

6.

```
$ 53.83    $ 27.95    $ 42.26    $ 42.72
+ 24.38    + 18.74    + 28.78    + 19.98
```

Name ______________________________

Adding two 4-digit numbers, two or more regrouping steps.

1.	\$ 27.74 + 17.69	\$ 33.55 + 26.67	\$ 51.57 + 17.68	\$ 43.65 + 33.65
2.	\$ 67.68 + 38.75	\$ 65.72 + 18.88	\$ 57.63 + 28.78	\$ 30.14 + 29.87
3.	\$ 57.14 + 46.87	\$ 58.14 + 49.89	\$ 67.38 + 27.98	\$ 37.69 + 23.54
4.	\$ 46.37 + 54.83	\$ 81.63 + 29.68	\$ 67.32 + 28.89	\$ 66.25 + 27.95
5.	\$ 48.53 + 23.68	\$ 63.75 + 54.25	\$ 91.54 + 19.76	\$ 48.89 + 35.31
6.	\$ 72.43 + 38.79	\$ 49.95 + 20.46	\$ 85.36 + 65.75	\$ 56.23 + 35.97

Name ______________________________

Adding three 3-digit numbers, two or more regrouping steps.

Example:

```
  $ 8.26
    3.72
  + 1.35
  $13.33
```

1.	$ 4.52 2.19 + 5.62	$ 3.15 6.83 + 4.24	$ 3.25 4.50 + 4.85	$ 3.84 6.24 + 2.16
2.	$ 4.21 3.52 + 4.49	$ 5.26 2.48 + 3.51	$ 4.19 2.98 + 5.16	$ 3.06 4.98 + 2.62
3.	$ 1.87 5.85 + 6.02	$ 3.62 2.98 + 5.42	$ 7.93 3.52 + 2.49	$ 8.52 2.41 + 5.09
4.	$ 7.19 1.99 + 4.22	$ 8.43 3.25 + 2.43	$ 6.10 3.77 + 1.36	$ 4.57 4.07 + 2.98
5.	$ 2.68 3.18 + 5.40	$ 3.53 2.97 + 5.03	$ 7.24 2.58 + 3.42	$ 4.25 5.25 + 6.56
6.	$ 4.10 2.99 + 3.17	$ 9.09 2.12 + 3.92	$ 9.21 3.67 + 2.34	$ 5.58 3.44 + 2.90

Name ______________________________

Adding three 3-digit numbers, two or more regrouping steps.

1.	\$ 3.46 6.25 + 4.52	\$ 6.37 2.49 + 3.14	\$ 7.92 2.25 + 1.47	\$ 4.29 5.82 + 2.91
2.	\$ 8.52 2.73 + 1.26	\$ 4.95 2.53 + 4.08	\$ 7.52 3.66 + 2.15	\$ 8.53 2.25 + 5.42
3.	\$ 7.24 1.86 + 6.24	\$ 8.18 2.73 + 3.42	\$ 8.36 5.41 + 1.66	\$ 3.98 2.49 + 5.21
4.	\$ 6.54 3.78 + 2.50	\$ 3.99 2.21 + 6.15	\$ 3.18 7.26 + 1.91	\$ 9.24 1.87 + 2.11
5.	\$ 6.06 5.43 + 2.81	\$ 3.64 2.98 + 3.00	\$ 2.89 5.21 + 3.06	\$ 8.98 1.24 + 1.68
6.	\$ 7.23 6.75 + 2.24	\$ 5.49 1.26 + 3.50	\$ 4.85 2.13 + 5.26	\$ 5.09 2.94 + 3.12

Name ______________________________

Adding three 4-digit numbers, two or more regrouping steps.

Example:

```
  $ 46.37
    19.42
  + 11.66
  $ 77.45
```

1.	$ 33.36 31.07 + 32.63	$ 32.22 51.67 + 13.28	$ 18.26 14.69 + 76.31	$ 76.14 12.82 + 23.15
2.	$ 61.08 11.63 + 28.17	$ 31.48 36.16 + 21.40	$ 22.36 66.46 + 11.24	$ 31.86 44.48 + 33.26
3.	$ 41.37 22.77 + 21.34	$ 30.27 26.68 + 47.06	$ 42.70 16.46 + 20.47	$ 60.83 18.55 + 21.14
4.	$ 64.82 27.37 + 14.37	$ 13.24 24.87 + 32.35	$ 37.04 56.34 + 45.86	$ 36.83 25.96 + 33.41
5.	$ 64.02 32.54 + 16.68	$ 32.70 25.09 + 26.46	$ 24.04 56.35 + 46.91	$ 36.81 27.94 + 34.36
6.	$ 44.57 13.38 + 66.37	$ 66.43 35.26 + 25.33	$ 47.24 26.02 + 32.88	$ 46.57 66.46 + 29.99

Name ________________________________

Adding three 4-digit numbers, two or more regrouping steps.

1.	\$ 48.36 62.53 + 22.36	\$ 20.68 24.32 + 18.26	\$ 42.06 14.63 + 35.42	\$ 63.73 62.58 + 12.92
2.	\$ 47.46 53.18 + 11.62	\$ 43.39 15.86 + 45.21	\$ 40.07 52.82 + 39.21	\$ 35.74 37.26 + 34.17
3.	\$ 92.49 16.25 + 13.71	\$ 25.68 48.12 + 63.32	\$ 43.62 42.85 + 24.18	\$ 37.25 47.82 + 52.13
4.	\$ 42.99 25.11 + 48.10	\$ 95.59 16.48 + 12.22	\$ 26.79 54.23 + 41.13	\$ 38.49 52.63 + 40.22
5.	\$ 64.87 95.43 + 47.93	\$ 89.68 71.12 + 60.89	\$ 77.29 63.25 + 51.87	\$ 94.95 83.27 + 56.25
6.	\$ 67.25 93.29 + 69.86	\$ 64.82 32.50 + 99.90	\$ 82.31 76.27 + 63.85	\$ 56.29 31.87 + 87.95

Name ______________________________

Placing the decimal point and dollar sign

When you add money, there are some important things to remember.

1. When you write money, use the dollar sign ($) and decimal point (.). Always be sure to have two numbers after the decimal point.

   6¢ is written like this ⟶ $.06

2. When you add and subtract money, be sure to keep your columns straight. Put the dollars under dollars and cents under cents.

Like this ⟶

```
  $ 15.13        $ 15.13
+    4.11      -    5.11
  $ 19.24        $ 10.02
```

Put the numbers below in a column and then add them.

1. $4.12, $1.09, $.36

2. $10.50, $.34, $5.34

3. $8.56, $.72, $29.14

4. $12.82, $.66, $1.86

Name ______________________________

Figuring the cost when buying more than one of the same item

Find the cost of the following items.

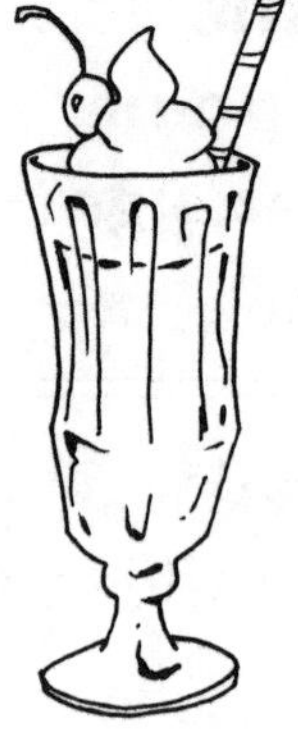

Example:
Two Cheese Omelettes @ $4.00 ea.

$ 4.00
+ 4.00
$ 8.00

1.	Two Jumbo Burgers	@	$5.95 ea.	________
2.	Two Shrimp Salads	@	$6.50 ea.	________
3.	Four Fish Dinners	@	$6.10 ea.	________
4.	Three pieces of Lemon Pie	@	$2.30 ea.	________
5.	Five Milk Shakes	@	$3.00 ea.	________
6.	Three Tuna Sandwiches	@	$3.95 ea.	________
7.	Four Fruit Salads	@	$4.95 ea.	________
8.	Three Hot Beef Sandwiches	@	$5.50 ea.	________
9.	Four Avocado Burgers	@	$5.75 ea.	________
10.	Five Club Sandwiches	@	$4.10 ea.	________
11.	Three Chicken Dinners	@	$6.50 ea.	________
12.	Four Reuben Sandwiches	@	$5.95 ea.	________
13.	Three Chef Salads	@	$5.65 ea.	________
14.	Five Bacon Burgers	@	$4.65 ea.	________

Name ______________________________

Figuring the cost when buying more than one of the same item

Use the menu to find the cost of the items. Write the total cost.

	ITEMS	ITEM COST	TOTAL
1.	Two Chili Burgers	$4.85	$9.70
2.	Two Spanish Omelettes		
3.	Three Tomato Salads		
4.	Four Tuna Sandwiches		
5.	Five Pieces of Lemon Pie		
6.	Four Colas		
7.	Five Waldorf Salads		
8.	Three Fruit Salads		
9.	Four Steak Dinners		
10.	Three orders of French Toast		
11.	Five cups of Hot Chocolate		
12.	Four Hot Beef Sandwiches		
13.	Three Hut Burgers		
14.	Two Patty Melts		

Name ______________________________

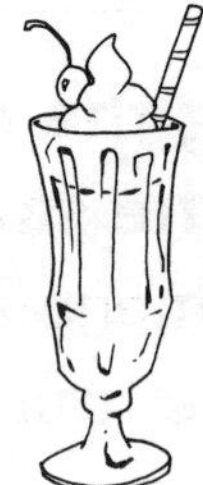

Solving addition story problems
Use the menu to solve the story problems (answers exclude sales tax).

1. What would be the cost of one Steak Dinner, a cup of Coffee, and a piece of Chocolate Cake?

   ____

2. How much would you have to pay for one Fish Dinner, one Cola, and a piece of Blueberry Pie?

   ____

3. What would be the cost of one Jumbo Burger and two Milk Shakes?

   ____

4. Find the cost of a bowl of Tomato Soup, a Cheese Sandwich, and a glass of Milk.

   ____

5. A Reuben Sandwich, a side order of Onion Rings, and a Cola would cost how much?

   ____

6. What must you pay for a bowl of Onion Soup, a Waldorf Salad, and two cups of Coffee?

   ____

7. How much would two Bacon Burgers, one Chili Burger, and two Colas cost?

   ____

8. Find the cost of a Shrimp Salad, a side order of Cole Slaw, a piece of Lemon Pie, and a cup of Coffee.

   ____

Name ______________________________

Solving addition story problems
Use the menu to solve the story problems (answers exclude sales tax).

1. What will you pay for a Spanish Omelette and a cup of Tea?

   ____

2. How much would a Chicken Sandwich, a bowl of Vegetable Soup, and a piece of Cherry Pie cost?

   ____

3. Find the cost of a Bacon Burger, a side order of Cottage Cheese, and a glass of Iced Tea.

   ____

4. How much would a Tomato Salad, Hot Tea, and a piece of Cheesecake cost?

   ____

5. What amount will you owe if you order Waffles, Orange Juice, and Coffee?

   ____

6. How much would you pay for Beef Stew, a Cola, and Lemon Pie?

   ____

7. What would your check total if you ordered a Hot Beef Sandwich and two side orders of Baked Beans?

   ____

8. Find the cost of a Chef Salad, a Meat Loaf Dinner, and a glass of Milk.

   ____

Name ______________________

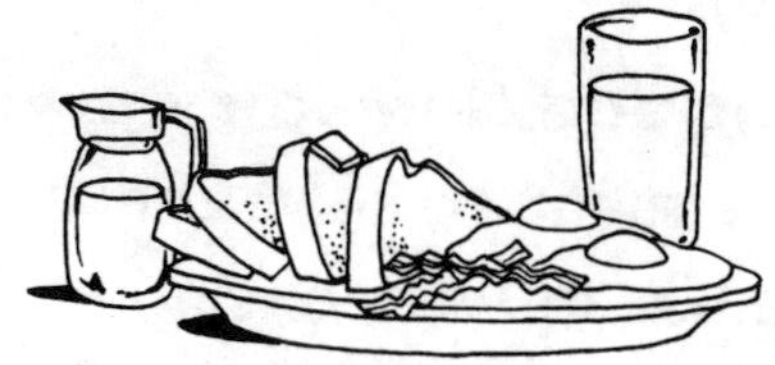

Solving addition story problems

Use the menu to solve the story problems (answers exclude sales tax).

1. David and John each order a Jumbo Burger and a cup of Hot Chocolate. How much do they owe?

 ____

2. Mr. Wilson stopped for breakfast on his way to work. He had Ham and Eggs, Grapefruit Juice, and Coffee. How much did his breakfast cost?

 ____

3. Sheila, Lisa, and Jim stopped at The Hamburger Hut for dinner. Lisa and Sheila shared a Hut Burger and a side order of Onion Rings. Jim ordered a Hot Turkey Sandwich and a piece of Cherry Pie. What amount did they owe?

 ____

4. Ted and Allison decided to have hamburgers after a movie. Ted had an Avocado Burger and a Milk Shake. Allison ordered the Patty Melt and Root Beer. What was the total of their bill?

 ____

5. After the concert, Tim and Rita went to The Hamburger Hut. Tim ordered a Bacon Burger and a Cola. Rita had a Spanish Omelette and Coffee. Tim paid the bill and left the waitress a $2.00 tip. How much did he spend?

 ____

6. You and your friend each order a Hut Burger, Cola, and Ice Cream. How much does the bill total?

 ____

Name ______________________________

Solving addition story problems

Use the menu to solve the story problems (answers exclude sales tax).

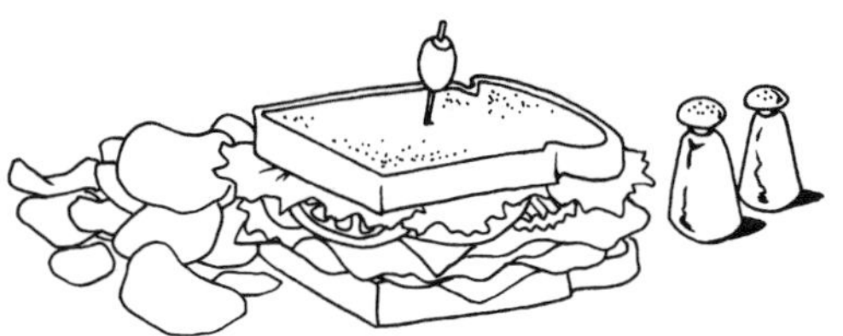

1. One Friday, Mrs. Wayne and Mrs. Lewis took their children to The Hamburger Hut for dinner. Mrs. Wayne had a Fish Dinner, Coffee, and Ice Cream. How much did her dinner cost?

   ____

2. Freddie ordered a Cheese Sandwich, a side order of Baked Beans, and a Milk Shake. What was the cost of Freddie's dinner?

   ____

3. Robert decided on a Hut Burger with cheese, a side order of Cole Slaw, Milk, and Cherry Pie. How much money did Robert spend on his dinner?

   ____

4. Beef Stew, Iced Tea, and Sherbet was what Mrs. Lewis had to eat. How much did Mrs. Lewis owe for her dinner?

   ____

5. Mike picked the Patty Melt, a side order of French Fries, and a Root Beer. He also ordered a piece of Banana Cream Pie for dessert. How much was dinner for Mike?

   ____

6. Mrs. Wayne spent $24.59 on dinner for her family and Mrs. Lewis spent $18.83. They left a $6.50 tip for the waitress. Altogether, how much did the two mothers spend for dinner at The Hamburger Hut?

   ____

Name ______________________________

Solving addition story problems

Use the menu to solve the story problems (answers exclude sales tax).

1. Carlo bought a Hut Burger, a side order of French Fries, and a glass of Milk. How much did he owe?

   ____

2. Nick took his girlfriend to The Hamburger Hut for breakfast. He ordered Bacon and Eggs and Orange Juice, and she had Cereal. How much did Nick pay for their breakfast?

   ____

3. The Women's Club had a dinner meeting at The Hamburger Hut. Mrs. Drake ordered the Beef Stew Dinner, Coffee, and Chocolate Cake. What did her check total?

   ____

4. Mrs. Wright ordered a Hot Beef Sandwich, Cheesecake, and Coffee. How much did Mrs. Wright owe?

   ____

5. Mrs. Francis was on a diet, so she ordered Fruit Salad and Hot Tea. What did her check total?

   ____

6. Mrs. Stone was very hungry. She decided to have a Steak Dinner, a side order of Onion Rings, Coffee, and Banana Cream Pie. How much was her check?

   ____

7. After looking at the menu, Mrs. Nelson ordered Liver and Onions, Root Beer, and Sherbet. What did her dinner cost?

   ____

Name ______________________________

Solving addition story problems
Use the menu to solve the story problems (answers exclude sales tax).

1. Sandy, Charlotte, and Dave each have some money to spend at The Hamburger Hut. Sandy decides to have an Egg Salad Sandwich, Root Beer, and a piece of Apple Pie. How much will this cost her? ________

2. Charlotte orders a Crab Salad, a side order of Onion Rings, and Milk. What will her check total? ________

3. Dave will order a Bacon Burger, a side order of Chili and Beans, and a Milk Shake. How much will he spend? ________

4. Willy wants a Hut Burger with a side order of French Fries and a Cola. What will he pay the cashier? ________

5. Betty and Pat stopped at The Hamburger Hut for dessert. Betty ordered Chocolate Cake and Coffee. Pat had Carrot Cake and a glass of Milk. How much did they spend in all? ________

6. Two secretaries go to The Hamburger Hut for lunch. One has a Reuben Sandwich and Iced Tea. The other secretary has a Club Sandwich and Lemon Pie. Altogether, what do they spend for lunch? ________

Name ______________________________

Solving addition story problems
Use the menu to solve the story problems (answers exclude sales tax).

1. After church, Mr. and Mrs. Meyers and their children stopped at The Hut for a late breakfast. Mr. Meyers had Bacon and Eggs, Tomato Juice, and Coffee. How much did his breakfast cost?

   ____

2. Mrs. Meyers chose French Toast, Orange Juice, and Hot Tea. How much was her breakfast?

   ____

3. Scott ordered Hot Cakes, Grapefruit Juice, and Milk. What did his meal cost?

   ____

4. Linda decided she was not very hungry. She had a side order of Toast and a glass of Milk. How much did Linda's food cost?

   ____

5. Lucy had Waffles, Tomato Juice, and Hot Chocolate. Find the cost of Lucy's breakfast.

   ____

6. The cost of breakfast for the Meyers family was $26.29. Mr. Meyers left a $4.50 tip. What was the total amount Mr. Meyers spent at The Hamburger Hut?

   ____

Name ______________________________

Solving addition story problems
Use the menu to solve the story problems (answers exclude sales tax).

1. On Saturday night, Dan took Brenda to The Hamburger Hut for dinner. Dan had a Steak Dinner, a glass of Milk, and Cherry Pie. Brenda had a Chicken Dinner and a Cola. What did their check total?

   ____

2. Stan, Ken, and Alan went out for breakfast. Stan ordered Waffles, Ken had Hot Cakes, and Alan ate French Toast. How much in all did the boys spend on breakfast?

   ____

3. Mrs. Marx treated her two sons to lunch. Steve had a Tuna Sandwich and Milk. Roger ordered a Jumbo Burger and a side order of Cole Slaw. Mrs. Marx had Beef Stew and Coffee. What did lunch cost for Mrs. Marx and her two sons?

   ____

4. Mr. Anthony stopped at The Hamburger Hut after work. He ordered hamburgers to take home to his family. He bought a Hut Burger, two Jumbo Burgers, and two Bacon Burgers. How much was his bill?

   ____

5. Laurie took her sister shopping and out for lunch. She ordered Shrimp Salad, Onion Rings, and a glass of Milk. Her sister had Fruit Salad, Cherry Pie, and a Cola. How much was spent on lunch?

   ____

6. Two accountants go to The Hamburger Hut for lunch. One has a Crab Salad and Iced Tea. The other accountant has a Club Sandwich and Hot Tea. Altogether, what do they spend for lunch?

   ____

Name ___________________________________

Solving addition story problems
Solve the story problems without the menu.

1. Katie works as a waitress at The Hamburger Hut. On Saturday, her tips totaled $18.55 and on Sunday, she made $19.20. In all, how much did Katie make on tips that weekend?

 ____

2. In one week, The Hut uses 60 loaves of white bread, 42 of wheat, and 16 of rye. How many loaves of bread does The Hamburger Hut use in one week?

 ____

3. The Golden Crust Bakery supplies pies to The Hamburger Hut. One summer, the following numbers of pies were delivered: 86 in June, 107 in July, and 119 in August. What was the total number of pies delivered to The Hamburger Hut?

 ____

4. Chuck was a busboy at The Hamburger Hut and could ride his bike to work. One day, he rode ten blocks to work, six blocks to the Post Office, and sixteen blocks back home. How far did he ride that day?

 ____

5. During the summer, Greg works as a dishwasher at The Hut. Last summer, he earned $1,206.15, and this summer he made $1,367.17. How much did Greg earn the last two summers?

 ____

6. One day at The Hamburger Hut, 20 pieces of Apple Pie, 36 pieces of Lemon Pie, and 54 pieces of Chocolate Cake were sold. Altogether, how many pieces of these desserts were sold?

 ____

Name ______________________________

Solving addition story problems

Solve the story problems without the menu.

1. Pete's Poultry Farm sells chickens to The Hamburger Hut. In September, Pete was paid $368.14; in October, $197.86; and in November, $322.17. How much money did The Hut spend on poultry those three months?

   ____

2. Richard was a dishwasher at The Hut. He was saving money for a trip to California. In March, he saved $58.30; in April, $63.57; and in May, $69.88. How much did he save in three months?

   ____

3. Joan and her twin sister, Janet, wait on tables at The Hamburger Hut. During January, Joan's paycheck totaled $410.86, and Janet made $429.37. Altogether, how much did the twins earn in January?

   ____

4. Dan worked as a cashier. He saved four of his paychecks for college. His checks amounted to $168.40, $147.90, $89.30, and $96.97. How much did he save?

   ____

5. Dave's Dairy delivers milk, eggs, and cheese to The Hut. So far this year, The Hut has paid Dave $289.30, $410.11, $381.50, $264.30, $85.89, and $129.17. How much money has been spent on dairy products?

   ____

6. Some repairs had to be made at The Hamburger Hut. A new door cost $983.00. Repairs to the floor came to a total of $787.60, and a cracked window was replaced for $305.10. What was the total cost of these repairs?

   ____

FINDING SALES TAX

Some states charge sales tax. When you buy food in a restaurant, you must pay sales tax. The amount of money you pay for sales tax depends on how much the food costs. Look carefully at the example below.

HAMBURGER HUT GUEST CHECK		
1	Jumbo Burger	$5.95
1	Root Beer	1.23
	SUBTOTAL	7.18
	(7% sales) *TAX*	.50
Thank You!	**TOTAL**	7.68

Follow the steps below:

1. Add the two items to find the **subtotal**.
2. Write in the **tax** from the Tax Table.
3. Add the **subtotal** and the **tax** to find the **total**.

7% SALES TAX TABLE

Amount of Sale	Tax	Amount of Sale	Tax
4.65 - 4.78	.33	6.36 - 6.49	.45
4.79 - 4.92	.34	6.50 - 6.64	.46
4.93 - 5.07	.35	6.65 - 6.78	.47
5.08 - 5.21	.36	6.79 - 6.94	.48
5.22 - 5.35	.37	6.95 - 7.07	.49
5.36 - 5.49	.38	**7.08 - 7.21**	**.50**
5.50 - 5.64	.39	7.22 - 7.35	.51
5.65 - 5.78	.40	7.36 - 7.49	.52
5.79 - 5.92	.41	7.50 - 7.64	.53
5.93 - 6.07	.42	7.65 - 7.78	.54
6.08 - 6.21	.43	7.79 - 7.92	.55
6.22 - 6.35	.44	7.93 - 8.07	.56

You will be practicing with a 7% Sales Tax Table.

Find out from your teacher what tax rate your state charges.

The **subtotal** of the sample check above is **$7.18**. **$7.18** is between **$7.08** and **$7.21** on the Tax Table. The tax owed is **$.50**.

7% SALES TAX TABLE

Amount of Sale	Tax	Amount of Sale	Tax	Amount of Sale	Tax
8.93 - 9.07	.63	14.93 - 15.07	1.05	20.93 - 21.07	1.47
9.08 - 9.21	.64	15.08 - 15.21	1.06	21.08 - 21.21	1.48
9.22 - 9.35	.65	15.22 - 15.35	1.07	21.22 - 21.35	1.49
9.36 - 9.49	.66	15.36 - 15.49	1.08	21.36 - 21.49	1.50
9.50 - 9.64	.67	15.50 - 15.64	1.09	21.50 - 21.64	1.51
9.65 - 9.78	.68	15.65 - 15.78	1.10	21.65 - 21.78	1.52
9.79 - 9.92	.69	15.79 - 15.92	1.11	21.79 - 21.92	1.53
9.93 - 10.07	.70	15.93 - 16.07	1.12	21.93 - 22.07	1.54
10.08 - 10.21	.71	16.08 - 16.21	1.13	22.08 - 22.21	1.55
10.22 - 10.35	.72	16.22 - 16.35	1.14	22.22 - 22.35	1.56
10.36 - 10.49	.73	16.36 - 16.49	1.15	22.36 - 22.49	1.57
10.50 - 10.64	.74	16.50 - 16.64	1.16	22.50 - 22.64	1.58
10.65 - 10.78	.75	16.65 - 16.78	1.17	22.65 - 22.78	1.59
10.79 - 10.92	.76	16.79 - 16.92	1.18	22.79 - 22.92	1.60
10.93 - 11.07	.77	16.93 - 17.07	1.19	22.93 - 23.07	1.61
11.08 - 11.21	.78	17.08 - 17.21	1.20	23.08 - 23.21	1.62
11.22 - 11.35	.79	17.22 - 17.35	1.21	23.22 - 23.35	1.63
11.36 - 11.49	.80	17.36 - 17.49	1.22	23.36 - 23.49	1.64
11.50 - 11.64	.81	17.50 - 17.64	1.23	23.50 - 23.64	1.65
11.65 - 11.78	.82	17.65 - 17.78	1.24	23.65 - 23.78	1.66
11.79 - 11.92	.83	17.79 - 17.92	1.25	23.79 - 23.92	1.67
11.93 - 12.07	.84	17.93 - 18.07	1.26	23.93 - 24.07	1.68
12.08 - 12.21	.85	18.08 - 18.21	1.27	24.08 - 24.21	1.69
12.22 - 12.35	.86	18.22 - 18.35	1.28	24.22 - 24.35	1.70
12.36 - 12.49	.87	18.36 - 18.49	1.29	24.36 - 24.49	1.71
12.50 - 12.64	.88	18.50 - 18.64	1.30	24.50 - 24.64	1.72
12.65 - 12.78	.89	18.65 - 18.78	1.31	24.65 - 24.78	1.73
12.79 - 12.92	.90	18.79 - 18.92	1.32	24.79 - 24.92	1.74
12.93 - 13.07	.91	18.93 - 19.07	1.33	24.93 - 25.07	1.75
13.08 - 13.21	.92	19.08 - 19.21	1.34	25.08 - 25.21	1.76
13.22 - 13.35	.93	19.22 - 19.35	1.35	25.22 - 25.35	1.77
13.36 - 13.49	.94	19.36 - 19.49	1.36	25.36 - 25.49	1.78
13.50 - 13.64	.95	19.50 - 19.64	1.37	25.50 - 25.64	1.79
13.65 - 13.78	.96	19.65 - 19.78	1.38	25.65 - 25.78	1.80
13.79 - 13.92	.97	19.79 - 19.92	1.39	25.79 - 25.92	1.81
13.93 - 14.07	.98	19.93 - 20.07	1.40	25.93 - 26.07	1.82
14.08 - 14.21	.99	20.08 - 20.21	1.41	26.08 - 26.21	1.83
14.22 - 14.35	1.00	20.22 - 20.35	1.42	26.22 - 26.35	1.84
14.36 - 14.49	1.01	20.36 - 20.49	1.43	26.36 - 26.49	1.85
14.50 - 14.64	1.02	20.50 - 20.64	1.44	26.50 - 26.64	1.86
14.65 - 14.78	1.03	20.65 - 20.78	1.45	26.65 - 26.78	1.87
14.79 - 14.92	1.04	20.79 - 20.92	1.46	26.79 - 26.92	1.88

Name ______________________________

Use the tax table on page 41 to find the sales tax on the following amounts.
Write the totals.

	Amount	Sales Tax	Total
1.	$8.99	________	________
2.	$9.56	________	________
3.	$10.37	________	________
4.	$11.00	________	________
5.	$12.05	________	________
6.	$14.80	________	________
7.	$15.09	________	________
8.	$16.67	________	________
9.	$18.25	________	________
10.	$19.46	________	________
11.	$22.83	________	________
12.	$24.78	________	________
13.	$25.50	________	________
14.	$26.53	________	________

Name ______________________________

1. Add the check to find the subtotal.
2. Use the tax table to find the amount of tax.
3. Add the subtotal and the tax to find the total.

HAMBURGER HUT
GUEST CHECK

1	Hut Burger	4.25
1	Avocado Burger	5.75
	SUBTOTAL	
	(7% sales) TAX	
Thank You!	TOTAL	

HAMBURGER HUT
GUEST CHECK

1	Shrimp Salad	6.50
1	Milk Shake	3.00
	SUBTOTAL	
	(7% sales) TAX	
Thank You!	TOTAL	

HAMBURGER HUT
GUEST CHECK

1	Steak Dinner	8.25
1	Coffee	.90
1	Cherry Pie	2.25
	SUBTOTAL	
	(7% sales) TAX	
Thank You!	TOTAL	

HAMBURGER HUT
GUEST CHECK

1	Cheese Omelette	4.00
1	French Toast	3.75
1	Coffee	.90
1	Orange Juice	1.04
	SUBTOTAL	
	(7% sales) TAX	
Thank You!	TOTAL	

Name ________________________________

1. Add the check to find the subtotal.
2. Use the tax table to find the amount of tax.
3. Add the subtotal and the tax to find the total.

HAMBURGER HUT GUEST CHECK		
1	Hot Turkey Sandwich	3.65
1	Chili and Beans	2.55
1	Cheesecake	3.00
	SUBTOTAL	
	(7% sales) ***TAX***	
Thank You!	**TOTAL**	

HAMBURGER HUT GUEST CHECK		
1	Spanish Omelette	4.80
1	French Toast	3.75
1	Cereal	3.00
	SUBTOTAL	
	(7% sales) ***TAX***	
Thank You!	**TOTAL**	

HAMBURGER HUT GUEST CHECK		
1	Reuben Sandwich	5.95
1	Milk Shake	3.00
1	Cheesecake	3.00
	SUBTOTAL	
	(7% sales) ***TAX***	
Thank You!	**TOTAL**	

HAMBURGER HUT GUEST CHECK		
1	Fish Dinner	6.10
1	Root Beer	1.23
1	Lemon Pie	2.30
	SUBTOTAL	
	(7% sales) ***TAX***	
Thank You!	**TOTAL**	

Name ______________________________

1. Add the check to find the subtotal.
2. Use the tax table to find the amount of tax.
3. Add the subtotal and the tax to find the total.

HAMBURGER HUT GUEST CHECK		
2	Reuben Sandwiches	11.90
2	Milks	2.20
1	Chocolate Cake	2.95
	SUBTOTAL	
	(7% sales) ***TAX***	
Thank You!	**TOTAL**	

HAMBURGER HUT GUEST CHECK		
2	Chili Burgers	9.70
2	Iced Teas	2.30
1	Carrot Cake	2.85
1	Ice Cream	1.65
	SUBTOTAL	
	(7% sales) ***TAX***	
Thank You!	**TOTAL**	

HAMBURGER HUT GUEST CHECK		
2	Crab Salads	14.50
2	Colas	2.50
1	Milk	1.10
1	Tomato Juice	1.04
	SUBTOTAL	
	(7% sales) ***TAX***	
Thank You!	**TOTAL**	

HAMBURGER HUT GUEST CHECK		
2	Plain Omelettes	6.80
1	Coffee	.90
1	Milk Shake	3.00
	SUBTOTAL	
	(7% sales) ***TAX***	
Thank You!	**TOTAL**	

Name ______________________________

1. Use the menu to find the prices.
2. Add the check to find the subtotal.
3. Use the tax table to find the amount of tax.
4. Add the subtotal and the tax.

HAMBURGER HUT
GUEST CHECK

2	Club Sandwiches	
1	Root Beer	
1	Cola	
2	Cherry Pies	
	SUBTOTAL	
	(7% sales) ***TAX***	
Thank You!	**TOTAL**	

HAMBURGER HUT
GUEST CHECK

1	Meat Loaf Dinner	
1	Crab Salad	
2	Coffees	
1	Banana Cream Pie	
1	Carrot Cake	
	SUBTOTAL	
	(7% sales) ***TAX***	
Thank You!	**TOTAL**	

HAMBURGER HUT
GUEST CHECK

2	Hot Cakes	
1	Orange Juice	
1	Tomato Juice	
1	Hot Tea	
1	Coffee	
	SUBTOTAL	
	(7% sales) ***TAX***	
Thank You!	**TOTAL**	

HAMBURGER HUT
GUEST CHECK

2	Waldorf Salads	
1	Chicken Sandwich	
1	Tuna Salad	
3	Colas	
	SUBTOTAL	
	(7% sales) ***TAX***	
Thank You!	**TOTAL**	

Name ______________________________

1. Use the menu to find the prices.
2. Add the check to find the subtotal.
3. Use the tax table to find the amount of tax.
4. Add the subtotal and the tax.

HAMBURGER HUT
GUEST CHECK

1	Vegetable Soup	
1	Clam Chowder Soup	
2	Cheese Sandwiches	
2	Milks	
	SUBTOTAL	
	(7% sales) ***TAX***	
Thank You!	**TOTAL**	

HAMBURGER HUT
GUEST CHECK

1	Cherry Pie	
1	Apple Pie	
2	Banana Cream Pies	
1	Coffee	
1	Hot Tea	
	SUBTOTAL	
	(7% sales) ***TAX***	
Thank You!	**TOTAL**	

HAMBURGER HUT
GUEST CHECK

2	Steak Dinners	
2	Hot Teas	
1	Chocolate Cake	
1	Coconut Cake	
	SUBTOTAL	
	(7% sales) ***TAX***	
Thank You!	**TOTAL**	

HAMBURGER HUT
GUEST CHECK

2	Cheese Omelettes	
1	Spanish Omelette	
1	Hot Chocolate	
1	Milk	
1	Orange Juice	
	SUBTOTAL	
	(7% sales) ***TAX***	
Thank You!	**TOTAL**	

Name ______________________________

1. Use the menu to find the prices.
2. Add the check to find the subtotal.
3. Use the tax table to find the amount of tax.
4. Add the subtotal and the tax.

HAMBURGER HUT GUEST CHECK		
2	Hut Burgers	
2	French Fries	
1	Cola	
1	Root Beer	
1	Ice Cream	
	SUBTOTAL	
	(7% sales) TAX	
Thank You!	TOTAL	

HAMBURGER HUT GUEST CHECK		
2	Fish Dinners	
2	Milks	
1	Cheesecake	
1	Blueberry Pie	
	SUBTOTAL	
	(7% sales) TAX	
Thank You!	TOTAL	

HAMBURGER HUT GUEST CHECK		
2	Chef Salads	
1	Tomato Soup	
1	Onion Soup	
1	Iced Tea	
1	Coffee	
	SUBTOTAL	
	(7% sales) TAX	
Thank You!	TOTAL	

HAMBURGER HUT GUEST CHECK		
3	Patty Melts	
2	Colas	
1	Root Beer	
1	Lemon Pie	
	SUBTOTAL	
	(7% sales) TAX	
Thank You!	TOTAL	

Name ______________________________

1. Use the menu to find the prices.
2. Add the check to find the subtotal.
3. Use the tax table to find the amount of tax.
4. Add the subtotal and the tax.

HAMBURGER HUT
GUEST CHECK

2	Reuben Sandwiches	
1	Milk Shake	
1	Orange Juice	
1	Onion Rings	
	SUBTOTAL	
	(7% sales) ***TAX***	
Thank You!	**TOTAL**	

HAMBURGER HUT
GUEST CHECK

1	Avocado Burger	
2	Jumbo Burgers	
4	Colas	
1	Pudding	
	SUBTOTAL	
	(7% sales) ***TAX***	
Thank You!	**TOTAL**	

HAMBURGER HUT
GUEST CHECK

2	Crab Salads	
2	Onion Soups	
1	Milk	
1	Coffee	
1	Apple Pie	
	SUBTOTAL	
	(7% sales) ***TAX***	
Thank You!	**TOTAL**	

HAMBURGER HUT
GUEST CHECK

2	Club Sandwiches	
2	Cheese Sandwiches	
1	Ham Sandwich	
2	Milk Shakes	
1	Root Beer	
	SUBTOTAL	
	(7% sales) ***TAX***	
Thank You!	**TOTAL**	

Name ____________________

1. Use the menu to find the prices.
2. Add the check to find the subtotal.
3. Use the tax table to find the amount of tax.
4. Add the subtotal and the tax.

HAMBURGER HUT
GUEST CHECK

2	Hot Beef Sandwiches	
1	Shrimp Salad	
1	Iced Tea	
1	Coffee	
2	Milks	
	SUBTOTAL	
	(7% sales) ***TAX***	
Thank You!	**TOTAL**	

HAMBURGER HUT
GUEST CHECK

3	Chili Burgers	
3	Colas	
1	Blueberry Pie	
1	Carrot Cake	
1	Lemon Pie	
	SUBTOTAL	
	(7% sales) ***TAX***	
Thank You!	**TOTAL**	

HAMBURGER HUT
GUEST CHECK

3	Beef Stew Dinners	
1	Egg Salad Sandwich	
2	Sherbets	
1	Ice Cream	
	SUBTOTAL	
	(7% sales) ***TAX***	
Thank You!	**TOTAL**	

HAMBURGER HUT
GUEST CHECK

2	Chef Salads	
2	Fruit Salads	
2	Coffees	
1	Hot Tea	
1	Blueberry Pies	
	SUBTOTAL	
	(7% sales) ***TAX***	
Thank You!	**TOTAL**	

Name ________________________________

Create your own checks.

HAMBURGER HUT GUEST CHECK		
	SUBTOTAL	
	(7% sales) ***TAX***	
Thank You!	**TOTAL**	

HAMBURGER HUT GUEST CHECK		
	SUBTOTAL	
	(7% sales) ***TAX***	
Thank You!	**TOTAL**	

HAMBURGER HUT GUEST CHECK		
	SUBTOTAL	
	(7% sales) ***TAX***	
Thank You!	**TOTAL**	

HAMBURGER HUT GUEST CHECK		
	SUBTOTAL	
	(7% sales) ***TAX***	
Thank You!	**TOTAL**	

Name ________________________________

POST-TEST – ADDITION

1. Write the amounts below using the dollar sign ($) and decimal point (.).

four cents	________
twenty-five cents	________
twelve dollars and three cents	________

2.

$ .43 + .31	$ .73 + .25	$ .33 .14 + .22	$ 1.22 + 2.34	$ 3.54 + 5.23

3.

$ 5.36 + 2.41	$ 3.23 4.32 + 1.43	$ 6.44 2.33 + 1.23	$ .48 + .23	$ .68 + .24

4.

$ .63 + .48	$ .86 + .36	$ .36 .74 + .21	$ 6.27 + 2.34	$ 5.48 + 3.25

5.

$ 5.56 + 2.75	$ 3.98 + 7.53	$ 7.98 + 2.36	$ 8.26 2.73 + 4.34

6.

$ 7.76 3.49 + 4.33	$ 46.63 + 23.48	$ 67.85 + 13.66	$ 23.62 46.38 + 35.23

Name ______________________________

POST-TEST – ADDITION

Solve the story problems
(answers exclude sales tax).

1. Marie ordered a Jumbo Burger for $5.95 and a Cola for $1.25. How much did she owe?

   ____

2. Dick and Dean had lunch at The Hamburger Hut. Dick had a Chicken Dinner for $6.50, a side order of Baked Beans for $.95, and a Milk Shake for $3.00. Dean decided on a Hut Burger for $4.25, Root Beer at $1.23, and a piece of Apple Pie for $2.30. How much did lunch for the two boys cost?

   ____

3. Find the cost of the following items:

Three Fruit Salads	@ $4.95 ea.	____________
Four Cheese Omelettes	@ $4.00 ea.	____________
Two Bacon Bugers	@ $4.65 ea.	____________
Three Fish Dinners	@ $6.10 ea.	____________

4. Shelly's paycheck for April was $211.90. In June, she earned $189.21, and in July, she made $214.68. How much did Shelly earn in those three months?

   ____

5. Jan, a waitress at The Hamburger Hut, is saving money for a car. She saved $168.52 in February, $192.12 in March, and $176.50 in April. How much did Jan save?

   ____

Name ________________________________

Subtracting two 2-digit numbers, no regrouping.

Example:

$ .48
- .16
$ .32

1.	$ .49 - .25	$ .96 - .76	$ .57 - .33	$ .99 - .29
2.	$ .48 - .31	$ .77 - .56	$ .88 - .24	$ .55 - .25
3.	$ .98 - .67	$ .36 - .25	$ .45 - .34	$ .77 - .16
4.	$ .28 - .16	$ .84 - .12	$ .97 - .67	$ .28 - .16
5.	$ .99 - .13	$ .42 - .31	$ .36 - .16	$ .69 - .32
6.	$ .77 - .33	$ .86 - .22	$ .58 - .14	$ .98 - .75

Name ___________________________________

Subtracting two 2-digit numbers, no regrouping.

1.	\$ .29 - .18	\$ .98 - .23	\$ .55 - .21	\$ .73 - .31
2.	\$ .85 - .54	\$ .36 - .25	\$ .48 - .31	\$ .37 - .15
3.	\$ .75 - .24	\$ .47 - .26	\$ .65 - .23	\$ .96 - .72
4.	\$ .28 - .17	\$ .59 - .48	\$ .39 - .23	\$ .96 - .45
5.	\$ .56 - .25	\$ .88 - .35	\$ .85 - .14	\$ .79 - .57
6.	\$ .76 - .23	\$ .63 - .12	\$ .57 - .35	\$ .49 - .36

Name ______________________________

Subtracting two 3-digit numbers, no regrouping.

Example:

$$\begin{array}{r} \$\ 8.36 \\ -\ \underline{3.14} \\ \$\ 5.22 \end{array}$$

---

1. $\begin{array}{r} \$\ 8.98 \\ -\ \underline{7.31} \end{array}$ $\begin{array}{r} \$\ 9.18 \\ -\ \underline{2.07} \end{array}$ $\begin{array}{r} \$\ 2.37 \\ -\ \underline{1.23} \end{array}$ $\begin{array}{r} \$\ 5.79 \\ -\ \underline{4.51} \end{array}$

2. $\begin{array}{r} \$\ 7.67 \\ -\ \underline{2.33} \end{array}$ $\begin{array}{r} \$\ 7.93 \\ -\ \underline{4.52} \end{array}$ $\begin{array}{r} \$\ 4.98 \\ -\ \underline{2.64} \end{array}$ $\begin{array}{r} \$\ 3.69 \\ -\ \underline{1.24} \end{array}$

3. $\begin{array}{r} \$\ 2.15 \\ -\ \underline{1.05} \end{array}$ $\begin{array}{r} \$\ 6.93 \\ -\ \underline{1.23} \end{array}$ $\begin{array}{r} \$\ 8.55 \\ -\ \underline{2.24} \end{array}$ $\begin{array}{r} \$\ 6.96 \\ -\ \underline{5.82} \end{array}$

4. $\begin{array}{r} \$\ 7.25 \\ -\ \underline{2.21} \end{array}$ $\begin{array}{r} \$\ 2.67 \\ -\ \underline{1.33} \end{array}$ $\begin{array}{r} \$\ 3.85 \\ -\ \underline{1.42} \end{array}$ $\begin{array}{r} \$\ 5.27 \\ -\ \underline{1.03} \end{array}$

5. $\begin{array}{r} \$\ 7.49 \\ -\ \underline{2.38} \end{array}$ $\begin{array}{r} \$\ 5.28 \\ -\ \underline{2.18} \end{array}$ $\begin{array}{r} \$\ 4.63 \\ -\ \underline{2.41} \end{array}$ $\begin{array}{r} \$\ 8.98 \\ -\ \underline{6.76} \end{array}$

6. $\begin{array}{r} \$\ 4.32 \\ -\ \underline{1.01} \end{array}$ $\begin{array}{r} \$\ 6.46 \\ -\ \underline{2.14} \end{array}$ $\begin{array}{r} \$\ 8.57 \\ -\ \underline{6.36} \end{array}$ $\begin{array}{r} \$\ 4.29 \\ -\ \underline{1.08} \end{array}$

Name ______________________________

Subtracting two 3-digit numbers, no regrouping.

1.	\$ 7.59 - 3.27	\$ 3.52 - 1.12	\$ 4.76 - 2.72	\$ 9.59 - 2.43
2.	\$ 7.65 - 4.31	\$ 9.98 - 2.57	\$ 8.69 - 1.49	\$ 5.46 - 2.35
3.	\$ 3.67 - 2.33	\$ 7.89 - 2.46	\$ 4.72 - 2.31	\$ 9.99 - 3.33
4.	\$ 6.38 - 1.23	\$ 4.98 - 1.65	\$ 3.68 - 1.24	\$ 9.56 - 4.55
5.	\$ 6.68 - 2.66	\$ 4.98 - 2.64	\$ 3.69 - 2.24	\$ 5.43 - 4.32
6.	\$ 6.83 - 4.61	\$ 5.25 - 2.03	\$ 8.57 - 3.14	\$ 4.69 - 1.23

Name ___________________________________

Subtracting two 4-digit numbers, no regrouping.

Example:
$ 18.57
- 12.23
$ 6.34

1.	$ 11.24 - 10.13	$ 14.06 - 12.02	$ 16.28 - 14.17	$ 24.05 - 12.04
2.	$ 23.16 - 13.13	$ 12.45 - 11.33	$ 36.04 - 25.02	$ 26.53 - 23.21
3.	$ 42.87 - 30.12	$ 60.56 - 50.13	$ 29.36 - 27.33	$ 45.99 - 41.85
4.	$ 17.74 - 14.12	$ 74.78 - 63.03	$ 13.69 - 11.34	$ 41.93 - 30.50
5.	$ 15.69 - 12.17	$ 53.84 - 41.63	$ 76.13 - 55.01	$ 53.99 - 22.43
6.	$ 36.29 - 14.18	$ 34.75 - 13.14	$ 18.97 - 16.24	$ 39.66 - 22.55

Name ____________________

Subtracting two 4-digit numbers, no regrouping.

1.	$ 95.37 - 12.12	$ 56.75 - 16.51	$ 77.28 - 64.11	$ 46.79 - 15.31
2.	$ 87.34 - 13.30	$ 79.46 - 62.11	$ 86.78 - 14.26	$ 36.25 - 13.23
3.	$ 75.36 - 63.22	$ 94.87 - 32.83	$ 67.89 - 11.22	$ 37.26 - 27.03
4.	$ 68.46 - 27.36	$ 58.85 - 47.24	$ 39.47 - 14.41	$ 45.35 - 35.05
5.	$ 84.89 - 63.63	$ 29.96 - 19.55	$ 79.98 - 26.53	$ 88.39 - 13.14
6.	$ 28.95 - 15.82	$ 93.98 - 73.64	$ 49.38 - 27.25	$ 94.87 - 82.40

Name ______________________________

Subtracting two 2-digit numbers, one-step regrouping.

Example:

$$\begin{array}{r} \$\ .\overset{3}{\not4}\overset{1}{2} \\ -\ \underline{.33} \\ \$\ .09 \end{array}$$

1.	$ .56 - .39	$ .80 - .75	$ .76 - .18	$ .55 - .38
2.	$ .46 - .37	$ .82 - .56	$ .32 - .16	$ .42 - .27
3.	$ .62 - .44	$ .52 - .35	$ .72 - .33	$ .83 - .66
4.	$ .83 - .17	$ .93 - .58	$ .76 - .48	$ .51 - .27
5.	$ .91 - .46	$ .45 - .19	$ .51 - .15	$ .85 - .27
6.	$ .66 - .57	$ .56 - .17	$ .68 - .39	$ .55 - .26

Name ________________________________

Subtracting two 2-digit numbers, one-step regrouping.

1.	\$ .84 - .25	\$ .63 - .54	\$ .56 - .27	\$ .47 - .38
2.	\$ .36 - .19	\$ .24 - .15	\$ .21 - .12	\$ .92 - .33
3.	\$ .84 - .27	\$ .64 - .55	\$ .50 - .31	\$ .42 - .23
4.	\$ .32 - .13	\$ .26 - .17	\$ .25 - .16	\$ .93 - .34
5.	\$ .86 - .29	\$ .68 - .49	\$ .52 - .14	\$ .47 - .38
6.	\$ .34 - .15	\$ .36 - .17	\$ .82 - .69	\$ .94 - .55

Name ___________________________________

Subtracting two 3-digit numbers, two-step regrouping.

Example:

	$ 3.37
-	1.68
	$ 1.69

1.	$ 3.21 - 1.36	$ 5.27 - 2.48	$ 7.26 - 1.48	$ 9.36 - 4.47
2.	$ 2.36 - 1.47	$ 4.15 - 2.26	$ 6.84 - 1.98	$ 8.35 - 2.49
3.	$ 3.46 - 1.59	$ 5.21 - 2.33	$ 7.21 - 4.34	$ 9.41 - 4.63
4.	$ 2.11 - 1.89	$ 4.63 - 2.74	$ 6.26 - 4.37	$ 8.31 - 6.66
5.	$ 3.42 - 1.98	$ 5.35 - 2.46	$ 7.47 - 1.58	$ 9.36 - 4.67
6.	$ 2.11 - 1.46	$ 4.32 - 2.43	$ 6.86 - 1.97	$ 8.30 - 5.55

Name ______________________________

Subtracting two 3-digit numbers, two-step regrouping.

1.	\$ 6.56 - 3.77	\$ 8.21 - 4.36	\$ 2.65 - 1.76	\$ 4.36 - 2.47
2.	\$ 3.15 - 2.98	\$ 5.25 - 1.48	\$ 7.26 - 3.57	\$ 9.22 - 3.77
3.	\$ 6.43 - 1.56	\$ 8.26 - 4.27	\$ 2.36 - 1.79	\$ 4.15 - 2.99
4.	\$ 3.66 - 1.77	\$ 5.42 - 3.53	\$ 7.83 - 4.94	\$ 9.63 - 4.76
5.	\$ 6.14 - 3.25	\$ 8.64 - 2.75	\$ 2.12 - 1.89	\$ 4.33 - 1.44
6.	\$ 3.12 - 1.67	\$ 5.22 - 1.45	\$ 7.26 - 5.67	\$ 9.36 - 4.27

Name ______________________________

Subtracting two 4-digit numbers, two or more regrouping steps.

Example:

$$\begin{array}{r} \$\ \overset{4}{\cancel{5}}\overset{16}{7}.\overset{15}{\cancel{6}}\overset{1}{\cancel{2}} \\ -\ \underline{38.73} \\ \$\ 18.89 \end{array}$$

1.	\$ 54.75 - 28.86	\$ 45.58 - 38.69	\$ 97.54 - 78.68	\$ 76.90 - 48.95
2.	\$ 63.91 - 54.96	\$ 95.70 - 67.85	\$ 87.63 - 59.78	\$ 63.98 - 54.99
3.	\$ 43.54 - 29.89	\$ 68.43 - 59.98	\$ 96.47 - 87.59	\$ 89.53 - 69.58
4.	\$ 75.97 - 38.99	\$ 95.94 - 56.85	\$ 85.43 - 26.85	\$ 99.43 - 78.47
5.	\$ 35.81 - 28.98	\$ 84.83 - 55.88	\$ 73.43 - 26.87	\$ 88.32 - 69.96
6.	\$ 43.57 - 34.68	\$ 96.43 - 77.55	\$ 95.35 - 47.99	\$ 39.77 - 29.89

Name ______________________________

Subtracting two 4-digit numbers, two or more regrouping steps.

1.	\$ 63.57 - 41.83	\$ 76.51 - 27.65	\$ 54.73 - 26.85	\$ 93.75 - 76.86
2.	\$ 34.85 - 28.89	\$ 47.33 - 28.44	\$ 42.43 - 23.54	\$ 91.53 - 22.96
3.	\$ 87.63 - 39.86	\$ 27.64 - 19.85	\$ 75.77 - 56.89	\$ 65.34 - 28.46
4.	\$ 95.36 - 26.48	\$ 54.54 - 25.65	\$ 53.62 - 34.83	\$ 44.56 - 16.87
5.	\$ 48.26 - 29.47	\$ 83.21 - 35.34	\$ 57.63 - 28.74	\$ 52.31 - 13.56
6.	\$ 84.53 - 36.64	\$ 57.25 - 29.87	\$ 62.11 - 15.78	\$ 73.22 - 38.66

Name ______________________________

Subtracting,
zero difficulty

Example:

$$\begin{array}{r} \$\ \overset{5}{\cancel{6}}.\overset{9}{\cancel{0}}\overset{10}{\cancel{0}} \\ -\ \ 3.82 \\ \hline \$\ 2.18 \end{array}$$

1.	\$ 6.00 - 2.92	\$ 3.00 - 2.89	\$ 7.00 - .89	\$ 7.00 - 3.56
2.	\$ 8.00 - 3.27	\$ 9.00 - 6.43	\$ 9.00 - 1.75	\$ 5.00 - 2.98
3.	\$ 4.00 - .89	\$ 6.00 - 2.39	\$ 8.00 - 2.91	\$ 2.00 - 1.47
4.	\$ 8.00 - 2.69	\$ 9.00 - 4.73	\$ 6.00 - .84	\$ 2.00 - .97
5.	\$ 7.00 - 2.58	\$ 9.00 - 1.26	\$ 5.00 - 3.99	\$ 3.00 - 1.22
6.	\$ 4.00 - 2.99	\$ 7.00 - 4.95	\$ 9.00 - 6.66	\$ 3.00 - .79

Name ______________________________

Subtracting,
zero difficulty

Example:

$$\begin{array}{r} \$\ 30.00 \\ -\ \ \underline{3.68} \\ \$\ 26.32 \end{array}$$

1.	\$ 50.00 - 3.46	\$ 30.00 - 2.79	\$ 70.00 - 2.55	\$ 10.00 - 5.55
2.	\$ 12.00 - 8.75	\$ 40.00 - 6.59	\$ 70.00 - 3.68	\$ 60.00 - 6.98
3.	\$ 80.00 - 1.87	\$ 30.00 - 2.29	\$ 40.00 - 4.65	\$ 30.00 - 2.51
4.	\$ 10.00 - 2.86	\$ 40.00 - 3.78	\$ 50.00 - 5.87	\$ 60.00 - 3.62
5.	\$ 70.00 - 4.13	\$ 50.00 - 1.35	\$ 20.00 - 1.88	\$ 40.00 - 2.11
6.	\$ 60.00 - 6.56	\$ 80.00 - 5.91	\$ 30.00 - 4.47	\$ 10.00 - 6.88

Name ____________________________

Subtracting,
zero difficulty

Example:

```
   1 9 9
$ 20.00
- 11.97
$  8.03
```

1.	\$ 30.00 - 14.28	\$ 50.00 - 13.62	\$ 70.00 - 27.85	\$ 90.00 - 14.96
2.	\$ 20.00 - 18.36	\$ 20.00 - 12.15	\$ 40.00 - 32.87	\$ 60.00 - 35.89
3.	\$ 80.00 - 64.86	\$ 30.00 - 23.17	\$ 50.00 - 36.43	\$ 70.00 - 47.82
4.	\$ 90.00 - 83.31	\$ 20.00 - 16.52	\$ 20.00 - 11.89	\$ 40.00 - 23.81
5.	\$ 60.00 - 41.71	\$ 80.00 - 78.81	\$ 30.00 - 12.98	\$ 50.00 - 15.75
6.	\$ 70.00 - 52.14	\$ 90.00 - 72.22	\$ 60.00 - 13.69	\$ 20.00 - 11.12

Name ______________________________

Subtracting, zero difficulty

1.	\$ 30.00 - 14.62	\$ 20.00 - 11.33	\$ 50.00 - 15.84	\$ 40.00 - 26.17
2.	\$ 60.00 - 21.16	\$ 80.00 - 52.19	\$ 90.00 - 36.85	\$ 20.00 - 15.42
3.	\$ 50.00 - 21.83	\$ 40.00 - 36.27	\$ 60.00 - 46.29	\$ 80.00 - 52.81
4.	\$ 90.00 - 64.92	\$ 20.00 - 14.83	\$ 30.00 - 17.74	\$ 60.00 - 16.26
5.	\$ 50.00 - 22.25	\$ 40.00 - 11.27	\$ 80.00 - 49.99	\$ 60.00 - 47.83
6.	\$ 80.00 - 27.89	\$ 90.00 - 46.32	\$ 20.00 - 15.87	\$ 30.00 - 16.83

Name ______________________________

Subtracting to make change
Write the amount of money you should receive in change.

	COST	PAID	CHANGE
1.	$ .16	$ .20	________
2.	$ .25	$ 1.00	________
3.	$ .85	$ 2.00	________
4.	$ .59	$ .64	________
5.	$ .63	$ 2.00	________
6.	$ 1.09	$ 5.00	________
7.	$ 3.36	$ 4.00	________
8.	$ 5.35	$ 10.00	________
9.	$ 2.59	$ 5.00	________
10.	$ 7.25	$ 8.50	________
11.	$ 13.80	$ 15.00	________
12.	$ 11.86	$ 13.50	________
13.	$ 47.29	$ 50.00	________
14.	$ 41.07	$ 45.00	________

Name ________________________________

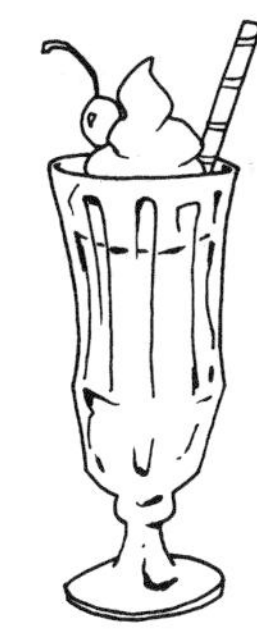

Using subtraction to find the difference in cost between two items

Use the menu. Find the difference in cost between the following items.

Example:

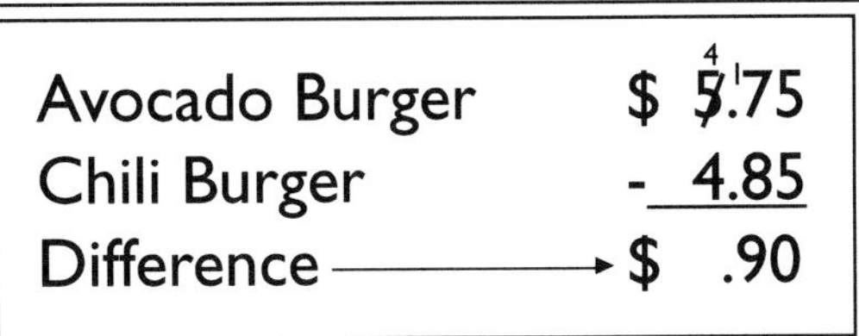

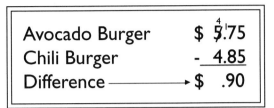

Avocado Burger	$ 5.75 (5 crossed out, 4 written above)
Chili Burger	- 4.85
Difference →	$ .90

Difference in Cost

1. A Ham Sandwich and a Chicken Sandwich ____________
2. Apple Pie and Cherry Pie ____________
3. Coconut Cake and Chocolate Cake ____________
4. A Spanish Omelette and a Plain Omelette ____________
5. A Hot Turkey Sandwich and a Reuben Sandwich ____________
6. A Shrimp Salad and a Crab Salad ____________
7. A Cola and a Milk Shake ____________
8. A Steak Dinner and a Fish Dinner ____________
9. A Cheese Sandwich and a Club Sandwich ____________
10. Pudding and Apple Pie ____________
11. A Reuben Sandwich and a Hut Burger ____________
12. Beef Stew and Meat Loaf ____________
13. A Waldorf Salad and a Tomato Salad ____________

Name ______________________________

Using subtraction to find out which item costs more

Use the menu to find the cost of each item. Subtract to find which item costs more.

Which Costs More?	Answer	How Much More?
1. French Fries or Onion Rings	French Fries	$.10
2. Cola or Root Beer		
3. Jumbo Burger or Chili Burger		
4. Vegetable Soup or Onion Soup		
5. Baked Beans or Chili and Beans		
6. Hot Cakes or Bacon and Eggs		
7. Lemon Pie or Sherbet		
8. Cheese Sandwich or Club Sandwich		
9. Liver and Onions or Meat Loaf		
10. Bacon Burger or Avocado Burger		
11. Cereal or Waffles		
12. Clam Chowder or Tomato Soup		
13. Juice or Milk Shake		
14. Cheesecake or Banana Cream Pie		

Name ______________________________

Solving subtraction story problems

Use the menu to solve the story problems (answers exclude sales tax).

1. How much more does an Avocado Burger cost than a Bacon Burger?

   ____

2. What is the difference in cost between a Meat Loaf Dinner and a Steak Dinner?

   ____

3. Bacon and Eggs cost more than French Toast. How much more?

   ____

4. What difference in cost is there between a Fish Dinner and a Chicken Dinner?

   ____

5. How much more must you pay for Clam Chowder than for Vegetable Soup?

   ____

6. You want to buy a Hut Burger with Cheese. You have only **$3.60** to spend. How much more do you need?

   ____

7. Which is more - a Jumbo Burger or a Bacon Burger? How much more?

   ____________ ____________

8. Find the difference in cost between a Waldorf Salad and a Tomato Salad.

   ____

Name ______________________________

Solving subtraction story problems
Use the menu to solve the story problems (answers exclude sales tax).

---

1. What difference in cost is there between a Hot Beef Sandwich and a Hot Turkey Sandwich?

   ____

2. How much more is Ham and Eggs than Bacon and Eggs?

   ____

3. The Crab Salad is more than the Shrimp Salad. How much more?

   ____

4. You want to order Hot Cakes and you have only $2.80. How much more will you need?

   ____

5. Find the difference in cost between Coconut Cake and Chocolate Cake.

   ____

6. How much more is Lemon Pie than Pudding?

   ____

7. If you have $3.75, how much more money will you need to buy a Jumbo Burger?

   ____

8. What is the difference in cost between a Chicken Dinner and a Steak Dinner?

   ____

Name ______________________________

Solving subtraction story problems

Solve the story problems without the menu.

---

1. The bill for dinner was $25.25. Chuck and Marty put their money together and found they had only $18.66. How much more will they need to pay the bill?

   ____

2. The manager found that there are only 28 cans of Orange Juice in the refrigerator. The Hamburger Hut usually uses 40 cans each week. How many more cans are needed?

   ____

3. Last year, The Hamburger Hut used 515 cases of ketchup. This year, the cook ordered 780 cases. How many more cases did The Hut order this year?

   ____

4. On Monday, Stan's Meat Market delivered 2,500 pounds of hamburger meat to The Hut. By Saturday, 509 pounds had been used. How many pounds of hamburger meat are left?

   ____

5. The chef figured he needs 2,000 hamburger buns. He saw that he has only 450 on hand. How many more buns will he have to order from the bakery?

   ____

6. Danny and Jeff are dishwashers at The Hut. Last week, Danny's paycheck totaled $130.80 and Jeff made $109.79. What was the difference in their paycheck totals?

   ____

Name ______________________________

Solving subtraction story problems

Solve the story problems without the menu.

1. Tammy saved all her tips one month. The total amount she saved was $210.55. She decided to spend $45.76 on CDs. How much of her tip money was left?

   ____

2. Sue Ellen, a cashier at The Hamburger Hut, is saving money for a trip to Hawaii in July. She needs $900.00. By April, she had saved $516.50. How much more will she have to save for her trip?

   ____

3. A car that Jason wants to buy costs $5,000.00. He has saved $3,269.00 by working weekends at The Hamburger Hut. How much more money does Jason need in order to have enough for the car?

   ____

4. Shelly, a waitress at The Hut, has put $1,296.25 in her savings account. Jennifer, Shelly's friend, has saved $1,364.50. Which girl has saved more money? How much more?

   ______________ ______________

5. Marge has been saving money. She wants to have $3,000 by the time school is out. She has already saved $1,865.55. How much more must she save?

   ____

6. Tina, the cook, earned $850.35 last month. She must pay $125 for child care. How much will she have left?

   ____

Name ______________________________

Using addition and subtraction to solve problems Use the menu to figure the costs. Subtract to find the change (exclude sales tax).

	You Buy	Cost	Pay with	Your Change
1.	Two Hut Burgers	________	$ 10.00	________
2.	Three Colas	________	$ 5.00	________
3.	Four pieces of Cheesecake	________	$ 15.00	________
4.	Two Steak Dinners	________	$ 20.00	________
5.	Five Crab Salads	________	$ 40.00	________
6.	Three Chicken Dinners	________	$ 20.00	________
7.	Four Spanish Omelettes	________	$ 50.00	________
8.	Six Root Beers	________	$ 8.00	________
9.	Two Chef Salads	________	$ 15.00	________
10.	Five Reuben Sandwiches	________	$ 40.00	________
11.	Three Fish Dinners	________	$ 20.00	________
12.	Four Jumbo Burgers	________	$ 50.00	________
13.	Six Avocado Burgers	________	$100.00	________
14.	Three Hot Beef Sandwiches	________	$ 50.00	________

Name ______________________________

Solving addition and subtraction story problems

Use the menu to solve the story problems (answers exclude sales tax).

---

1. Suppose that you buy a Hut Burger and a Cola. You pay for your food with a $10.00 bill. How much change should you receive?

   ____

2. If you order Bacon and Eggs, Orange Juice, and Milk and pay with a $10.00 bill, what amount of change will you get back?

   ____

3. Your friend has a Cheese Sandwich, Tomato Soup, and Milk. He gives the cashier $20.00. What amount of change should the cashier give him?

   ____

4. An order of Onion Rings, a Milk Shake, and a piece of Cherry Pie would cost how much? How much change would you get from $10.00?

   ____________ ____________

5. You would like to have a Steak Dinner, Iced Tea, and Chocolate Cake. You have $11.31. How much more money will you need?

   ____

6. What amount of money will you need to buy a Patty Melt, a Root Beer, and a piece of Carrot Cake? How much change from a $20.00 bill will the cashier give you?

   ____________ ____________

Name ______________________________

Solving addition and subtraction story problems

Use the menu to solve the story problems (answers exclude sales tax).

---

1. Janet's mother gave her money to take her cousin to dinner. Janet orders Beef Stew, Milk, and Pudding. How much does her dinner cost?

   ____

2. Janet's cousin Ann has Meat Loaf, Iced Tea, and a piece of Lemon Pie. How much will be spent for Ann's dinner?

   ____

3. When the two girls were finished eating, the waiter gave them their check, which totaled $17.80. How much change will Janet get back from $20.00?

   ____

4. After ordering two Hut Burgers, one side order of French Fries, and a Cola, Sam counted the money in his pocket. He found he has only $9.79. How much more money does he need to pay the bill?

   ____

5. Doug orders a Chili Burger, a side order of Onion Rings, and Root Beer. He discovers he has only $7.89. His brother will loan him enough money for the bill. How much must Doug borrow?

   ____

6. Stephanie is going to eat dinner at The Hut. She can't decide whether to order a Jumbo Burger or a Club Sandwich. How much money will she save by having a Club Sandwich?

   ____

Name ______________________________

Solving addition and subtraction story problems

Use the menu to solve the story problems (answers exclude sales tax).

---

1. Gina and four of her friends went to The Hamburger Hut for lunch. Gina had a Hut Burger, a side order of Baked Beans, and a Milk Shake. She paid for her lunch with a $10.00 bill. What amount of change will the cashier give her?

   ____

2. Rose decided on a Tuna Sandwich and a Root Beer. She gave the cashier $6.00. How much change should she receive?

   ____

3. Angela ordered a Jumbo Burger, a side order of Onion Rings, a Root Beer, and a piece of Banana Cream Pie. To pay for her lunch, she used a $20.00 bill. How much change will she get back?

   ____

4. For lunch, Katrina picked a Club Sandwich, a bowl of Tomato Soup, and Iced Tea. After paying for her lunch with a $50.00 bill, how much money will she have left?

   ____

5. Holly wanted to have an Avocado Burger and a glass of Milk. She had only $5.80. How much more money will she need?

   ____

6. The girls wanted to leave a $5.00 tip. They put their coins together and discovered it came to only $3.55. How much more money do they need for the tip?

   ____

Name ______________________________

Solving addition and subtraction story problems

Use the menu to solve the story problems (answers exclude sales tax).

1. You and your friend have lunch at The Hamburger Hut. You order a Tuna Sandwich and a glass of Milk. Your friend has a Bacon Burger and a Milk Shake. If you pay for both lunches with a $20.00 bill, how much change will you get back?

   ____

2. Imagine that you take $15.00 to The Hamburger Hut. You order a Chili Burger, a side order of Onion Rings, a Milk Shake, and a piece of Chocolate Cake. How much money will you have left?

   ____

3. Loren treated her Aunt Sara to dessert. Loren had Banana Cream Pie and Hot Chocolate. Sara ordered Cherry Pie and Coffee. What amount of change did Loren receive from $10.00?

   ____

4. Mr. Stevens took his daughters out for dinner. He had a Steak Dinner and Coffee. Betty ate a Patty Melt and a piece of Lemon Pie. Karen ordered a Jumbo Burger and a Milk Shake. How much money will Mr. Stevens have left from the $50.00 he uses to pay for the dinner?

   ____

5. Craig had $25.00 to take his girlfriend out for lunch. She ordered a Ham Sandwich and a Root Beer. He had a Hut Burger and a Cola. How much money does Craig have left?

   ____

Name ___________________________________

Adding to find total cost and subtracting to find change received

After the school dance, several couples went to The Hamburger Hut to eat. Many of them sat together but placed their orders separately.

Find the total of each order and the amount of change received. Use the menu to answer the problems. (All answers exclude sales tax.)

ORDER	COST	PAY	CHANGE
1. 2 Hut Burgers		$20.00	
2 Cherry Pies			
2 Colas			
Total			
2. 1 Jumbo Burger		$15.00	
1 Avocado Burger			
2 Iced Teas			
Total			
3. 2 Steak Dinners		$30.00	
2 Coconut Cakes			
3 Coffees			
Total			
4. 2 Patty Melts		$50.00	
3 Root Beers			
2 Cheesecakes			
1 Fish Dinner			
1 Apple Pie			
Total			

Name ______________________________

1. Add the check to find the subtotal.
2. Use the tax table to find the amount of tax.
3. Add the subtotal and the tax.
4. Subtract the total from the amount paid to find the change received.

HAMBURGER HUT
GUEST CHECK

1	Jumbo Burger	5.95
1	Milk Shake	3.00
1	Onion Rings	2.10
1	Chocolate Cake	2.95
	SUBTOTAL	
	(7% sales) **TAX**	
Thank You!	**TOTAL**	

Amount Paid $20.00 ____________ change

HAMBURGER HUT
GUEST CHECK

2	Chili Burgers	9.70
1	French Fries	2.20
1	Milk	1.10
	SUBTOTAL	
	(7% sales) **TAX**	
Thank You!	**TOTAL**	

Amount Paid $15.00 ____________ change

HAMBURGER HUT
GUEST CHECK

1	Fish Dinner	6.10
1	Coffee	.90
1	Cherry Pie	2.25
	SUBTOTAL	
	(7% sales) **TAX**	
Thank You!	**TOTAL**	

Amount Paid $10.00 ____________ change

HAMBURGER HUT
GUEST CHECK

1	Spanish Omelette	4.80
1	Plain Omelette	3.40
2	Hot Teas	2.30
	SUBTOTAL	
	(7% sales) **TAX**	
Thank You!	**TOTAL**	

Amount Paid $12.00 ____________ change

Name ______________________________

1. Add the check to find the subtotal.
2. Use the tax table to find the amount of tax.
3. Add the subtotal and the tax.
4. Subtract the total from the amount paid to find the change received.

HAMBURGER HUT GUEST CHECK		
1	Chicken Sandwich	4.25
1	Ham Sandwich	4.50
2	Tomato Soups	3.00
2	Milk Shakes	6.00
	SUBTOTAL	
	(7% sales) ***TAX***	
Thank You!	**TOTAL**	

Amount Paid $20.00 ____________ change

HAMBURGER HUT GUEST CHECK		
1	Shrimp Salad	6.50
1	Tomato Salad	5.25
1	Coffee	.90
2	Milks	2.20
	SUBTOTAL	
	(7% sales) ***TAX***	
Thank You!	**TOTAL**	

Amount Paid $20.00 ____________ change

HAMBURGER HUT GUEST CHECK		
1	Club Sandwich	4.10
1	Tuna Sandwich	3.95
1	Root Beer	1.23
1	Iced Tea	1.15
	SUBTOTAL	
	(7% sales) ***TAX***	
Thank You!	**TOTAL**	

Amount Paid $15.00 ____________ change

HAMBURGER HUT GUEST CHECK		
1	Bacon Burger	4.65
1	Cheese Omelette	4.00
1	Coffee	.90
1	Milk	1.10
	SUBTOTAL	
	(7% sales) ***TAX***	
Thank You!	**TOTAL**	

Amount Paid $20.00 ____________ change

Name ________________________________

1. Add the check to find the subtotal.
2. Use the tax table to find the amount of tax.
3. Add the subtotal and the tax.
4. Subtract the total from the amount paid to find the change received.

HAMBURGER HUT
GUEST CHECK

1	Steak Dinner	8.25
1	Meat Loaf Dinner	5.80
2	Iced Teas	2.30
	SUBTOTAL	
	(7% sales) ***TAX***	
Thank You!	**TOTAL**	

Amount Paid $50.00 ____________ change

HAMBURGER HUT
GUEST CHECK

2	Cheese Sandwiches	6.00
2	Tuna Sandwiches	7.90
2	Colas	2.50
2	Root Beers	2.46
	SUBTOTAL	
	(7% sales) ***TAX***	
Thank You!	**TOTAL**	

Amount Paid $25.00 ____________ change

HAMBURGER HUT
GUEST CHECK

1	Cherry Pie	2.25
1	Chocolate Cake	2.95
1	Sherbet	1.45
3	Coffees	2.70
	SUBTOTAL	
	(7% sales) ***TAX***	
Thank You!	**TOTAL**	

Amount Paid $20.00 ____________ change

HAMBURGER HUT
GUEST CHECK

2	orders Waffles	7.70
1	Plain Omelette	3.40
1	Milk	1.10
1	Orange Juice	1.04
	SUBTOTAL	
	(7% sales) ***TAX***	
Thank You!	**TOTAL**	

Amount Paid $20.00 ____________ change

Name ______________________________

1. Add the check to find the subtotal.
2. Use the tax table to find the amount of tax.
3. Add the subtotal and the tax.
4. Subtract the total from the amount paid to find the change received.

HAMBURGER HUT
GUEST CHECK

2	Bacon & Eggs	8.90
1	Ham & Eggs	4.95
1	Coffee	.90
2	Orange Juices	2.08
	SUBTOTAL	
	(7% sales) ***TAX***	
Thank You!	**TOTAL**	

Amount Paid $20.00 ____________ change

HAMBURGER HUT
GUEST CHECK

1	Chef Salad	5.65
1	Steak Dinner	8.25
1	Cheesecake	3.00
	SUBTOTAL	
	(7% sales) ***TAX***	
Thank You!	**TOTAL**	

Amount Paid $19.00 ____________ change

HAMBURGER HUT
GUEST CHECK

1	Vegetable Soup	1.35
1	Onion Soup	1.25
2	Hut Burgers	8.50
2	Milks	2.20
	SUBTOTAL	
	(7% sales) ***TAX***	
Thank You!	**TOTAL**	

Amount Paid $15.00 ____________ change

HAMBURGER HUT
GUEST CHECK

2	Chili Burgers	9.70
1	Patty Melt	4.55
1	Milk Shake	3.00
2	Colas	2.50
	SUBTOTAL	
	(7% sales) ***TAX***	
Thank You!	**TOTAL**	

Amount Paid $50.00 ____________ change

Name ______________________________

1. Use the menu to find the prices.
2. Add the check to find the subtotal.
3. Use the tax table to find the amount of tax.
4. Add the subtotal and the tax.
5. Subtract the total from the amount paid to find the change received.

HAMBURGER HUT
GUEST CHECK

1	Crab Salad	
1	Reuben Sandwich	
1	Milk	
1	Hot Tea	
2	Ice Creams	
	SUBTOTAL	
	(7% sales) ***TAX***	
Thank You!	**TOTAL**	

Amount Paid $25.00 __________ change

HAMBURGER HUT
GUEST CHECK

1	Club Sandwich	
1	Patty Melt	
1	Tomato Salad	
3	Coffees	
	SUBTOTAL	
	(7% sales) ***TAX***	
Thank You!	**TOTAL**	

Amount Paid $20.00 __________ change

HAMBURGER HUT
GUEST CHECK

1	French Toast	
1	Cereal	
1	Hot Chocolate	
2	Tomato Juice	
	SUBTOTAL	
	(7% sales) ***TAX***	
Thank You!	**TOTAL**	

Amount Paid $15.00 __________ change

HAMBURGER HUT
GUEST CHECK

2	Bacon Burgers	
1	Avocado Burger	
2	Colas	
1	Cherry Pie	
1	Apple Pie	
	SUBTOTAL	
	(7% sales) ***TAX***	
Thank You!	**TOTAL**	

Amount Paid $30.00 __________ change

Name ___________________________________

1. Use the menu to find the prices.
2. Add the check to find the subtotal.
3. Use the tax table to find the amount of tax.
4. Add the subtotal and the tax.
5. Subtract the total from the amount paid to find the change received.

HAMBURGER HUT
GUEST CHECK

1	Waldorf Salad	
1	Meat Loaf Dinner	
1	Steak Dinner	
3	Hot Teas	
2	Blueberry Pies	
	SUBTOTAL	
	(7% sales) ***TAX***	
Thank You!	**TOTAL**	

Amount Paid $30.00 ____________ change

HAMBURGER HUT
GUEST CHECK

1	Meat Loaf Dinner	
2	Chicken Dinners	
1	Waldorf Salad	
4	Coffees	
	SUBTOTAL	
	(7% sales) ***TAX***	
Thank You!	**TOTAL**	

Amount Paid $50.00 ____________ change

HAMBURGER HUT
GUEST CHECK

2	Cheese Sandwiches	
3	Ham Sandwiches	
2	Colas	
3	Iced Teas	
	SUBTOTAL	
	(7% sales) ***TAX***	
Thank You!	**TOTAL**	

Amount Paid $30.00 ____________ change

HAMBURGER HUT
GUEST CHECK

2	Hut Burgers	
1	Bacon Burger	
1	Patty Melt	
3	Milks	
1	Hot Chocolate	
	SUBTOTAL	
	(7% sales) ***TAX***	
Thank You!	**TOTAL**	

Amount Paid $40.00 ____________ change

Name ______________________________

1. Use the menu to find the prices.
2. Add the check to find the subtotal.
3. Use the tax table to find the amount of tax.
4. Add the subtotal and the tax.
5. Subtract the total from the amount paid to find the change received.

HAMBURGER HUT GUEST CHECK		
1	Chicken Sandwich	
2	Avocado Burgers	
2	Cole Slaws	
3	Iced Teas	
3	Ice Creams	
	SUBTOTAL	
	(7% sales) ***TAX***	
Thank You!	**TOTAL**	

Amount Paid $50.00 ____________ change

HAMBURGER HUT GUEST CHECK		
1	Chicken Dinner	
2	Chef Salads	
1	Waldorf Salad	
3	Root Beers	
1	Milk	
	SUBTOTAL	
	(7% sales) ***TAX***	
Thank You!	**TOTAL**	

Amount Paid $30.00 ____________ change

HAMBURGER HUT GUEST CHECK		
2	Avocado Burgers	
2	Hut Burgers	
5	Root Beers	
	SUBTOTAL	
	(7% sales) ***TAX***	
Thank You!	**TOTAL**	

Amount Paid $40.00 ____________ change

HAMBURGER HUT GUEST CHECK		
2	Spanish Omelettes	
3	Cereals	
4	Tomato Juices	
1	Orange Juice	
	SUBTOTAL	
	(7% sales) ***TAX***	
Thank You!	**TOTAL**	

Amount Paid $100.00 ____________ change

Name ___________________________________

1. Use the menu to find the prices.
2. Add the check to find the subtotal.
3. Use the tax table to find the amount of tax.
4. Add the subtotal and the tax.
5. Subtract the total from the amount paid to find the change received.

HAMBURGER HUT
GUEST CHECK

2	Patty Melts	
2	Reuben Sandwiches	
2	Cottage Cheeses	
3	Milks	
	SUBTOTAL	
	(7% sales) TAX	
Thank You!	TOTAL	

Amount Paid $30.00 __________ change

HAMBURGER HUT
GUEST CHECK

3	Cereals	
2	Ham & Eggs	
4	Grapefruit Juices	
3	Coffees	
	SUBTOTAL	
	(7% sales) TAX	
Thank You!	TOTAL	

Amount Paid $40.00 __________ change

HAMBURGER HUT
GUEST CHECK

3	Chicken Sandwiches	
2	Onion Soups	
3	Milk Shakes	
	SUBTOTAL	
	(7% sales) TAX	
Thank You!	TOTAL	

Amount Paid $40.00 __________ change

HAMBURGER HUT
GUEST CHECK

3	Hut Burgers	
2	Chili & Beans	
2	Cole Slaws	
5	Root Beers	
	SUBTOTAL	
	(7% sales) TAX	
Thank You!	TOTAL	

Amount Paid $100.00 __________ change

Name ______________________________

POST-TEST – Addition and Subtraction

1.	\$ .43 + .31	\$ .32 .24 + .21	\$ 3.24 + 1.43	\$ 4.45 3.32 + 1.20
2.	\$ .26 + .64	\$ 3.46 + 3.28	\$ 3.79 + 5.42	\$ 8.77 + 6.59
3.	\$ 8.75 2.84 + 4.27	\$ 56.99 + 27.94	\$ 53.87 26.35 + 12.35	\$ .57 - .24
4.	\$ 4.76 - 3.26	\$ 35.29 - 24.03	\$ .57 - .28	\$ 5.72 - 1.85
5.	\$ 26.38 - 19.49	\$ 5.00 - 2.97	\$ 40.00 - 8.76	\$ 70.00 - 56.43
6.	\$ 48.54 - 17.45	\$ 80.00 - 22.37	\$ 72.75 - 32.86	\$ 6.00 - 2.88

Answer Key

PG	
1	**1)** 5.95 **2)** .70 **3)** 5.80 **4)** 4.25 **5)** whipped **6)** 3.10 **7)** hashbrowns, toast + jelly **8)** Cheddar **9)** 3.80 **10)** French Fries **11)** 2.25 **12)** Orange, Tomato, & Grapefruit
2	**1)** .06 **2)** .08 **3)** .10 **4)** .15 **5)** .43 **6)** 1.25 **7)** 4.13 **8)** 14.05 **9)** 63.20 **10)** 84.06
3	**1)** .16 .19 .39 .22 **2)** .58 .39 .66 .48 **3)** .38 .39 .59 .85 **4)** .65 .47 .77 .69 **5)** .49 .68 .88 .79 **6)** .56 .86 .99 .89
4	**1)** .17 .19 .26 .28 **2)** .65 .38 .58 .77 **3)** .88 .46 .67 .88 **4)** .48 .79 .78 .69 **5)** .48 .98 .96 .89 **6)** .97 .98 .68 .68
5	**1)** .76 .99 .69 .67 **2)** .76 .99 .77 .74 **3)** .79 .98 .86 .67 **4)** .59 .48 .99 .99 **5)** .79 .89 .86 .89 **6)** .97 .99 .69 .89
6	**1)** .49 .98 .98 .89 **2)** .78 .59 .77 .79 **3)** .59 .79 .89 .85 **4)** .88 .79 .98 .79 **5)** .99 .79 .89 .99 **6)** .98 .97 .99 .99
7	**1)** 3.45 4.67 6.19 5.54 **2)** 5.19 4.39 8.17 5.28 **3)** 9.54 6.28 7.08 9.24 **4)** 7.85 9.49 7.78 7.59 **5)** 9.34 8.06 8.77 5.95 **6)** 9.18 7.87 9.56 9.39
8	**1)** 5.79 8.68 8.98 8.89 **2)** 7.97 9.68 9.77 9.98 **3)** 7.98 7.99 8.98 8.78 **4)** 9.99 8.78 9.99 9.86 **5)** 8.87 7.89 7.98 9.99 **6)** 8.67 9.88 9.99 9.68
9	**1)** 9.84 7.97 9.37 9.75 **2)** 7.38 8.85 8.78 8.57 **3)** 4.59 8.58 6.86 8.38 **4)** 8.88 8.98 8.57 6.79 **5)** 7.67 9.45 7.26 9.29 **6)** 8.79 7.97 8.87 9.18
10	**1)** 6.93 5.69 6.56 7.99 **2)** 6.39 7.89 7.67 9.64 **3)** 8.49 6.94 6.57 8.75 **4)** 5.54 8.96 9.85 5.37 **5)** 5.59 8.74 8.18 4.99 **6)** 6.69 8.38 8.57 6.19
11	**1)** .40 .60 .81 .63 **2)** .60 .41 .52 .87 **3)** .42 .41 .83 .52 **4)** .56 .90 .80 .61 **5)** .45 .64 .82 .50 **6)** .91 .90 .72 .91
12	**1)** .61 .41 .50 .52 **2)** .80 .50 .64 .92 **3)** .51 .52 .92 .63 **4)** .51 .42 .91 .84 **5)** .70 .81 .92 .93 **6)** .94 .45 .71 .64
13	**1)** 1.30 1.20 1.30 1.11 **2)** 1.21 1.72 1.72 1.22 **3)** 1.40 1.50 1.30 1.32 **4)** 1.46 1.74 1.51 1.76 **5)** 1.51 1.23 1.62 1.23 **6)** 1.12 1.02 1.33 1.92
14	**1)** 1.33 1.62 1.48 1.22 **2)** 1.44 1.74 1.43 1.98 **3)** 1.11 1.62 1.46 1.21 **4)** 1.04 1.43 1.17 1.36 **5)** 1.57 1.41 1.52 1.41 **6)** 1.28 1.43 1.20 1.13
15	**1)** 8.61 2.73 5.73 8.90 **2)** 7.62 3.96 8.72 2.60 **3)** 6.90 7.84 5.84 3.76 **4)** 5.90 2.50 4.66 7.92 **5)** 5.35 8.86 4.81 6.57 **6)** 9.21 9.60 9.58 8.85
16	**1)** 7.91 9.60 3.86 3.68 **2)** 9.85 2.73 3.95 7.90 **3)** 6.91 3.73 8.40 8.90 **4)** 8.82 5.91 9.61 7.86 **5)** 6.80 7.90 7.53 7.71 **6)** 8.75 8.92 7.70 6.51
17	**1)** 9.23 8.52 9.12 8.52 **2)** 9.42 4.00 6.81 9.81 **3)** 8.21 9.00 7.06 9.50 **4)** 7.32 8.34 4.00 8.05 **5)** 6.25 5.00 7.72 7.55 **6)** 9.21 9.00 7.10 9.48
18	**1)** 9.63 6.03 7.00 9.35 **2)** 7.10 6.76 5.22 3.43 **3)** 7.21 9.10 7.06 8.24 **4)** 7.04 9.22 4.58 5.00 **5)** 8.00 9.00 9.00 4.33 **6)** 8.82 4.30 9.82 6.82

PG	
19	**1)** 10.12 14.52 16.00 14.75 **2)** 12.25 11.45 15.33 14.22 **3)** 13.47 12.41 16.00 16.31 **4)** 13.16 13.30 12.12 13.36 **5)** 12.22 11.01 14.10 16.23 **6)** 15.74 11.37 14.50 13.33
20	**1)** 10.42 15.64 16.13 12.24 **2)** 11.18 11.33 12.25 11.43 **3)** 19.41 16.04 10.80 12.21 **4)** 12.13 14.50 15.36. 14.31 **5)** 11.04 14.33 10.27 11.20 **6)** 17.35 10.21 13.15 12.10
21	**1)** 38.31 59.30 86.00 51.03 **2)** 54.20 35.00 50.22 80.42 **3)** 90.11 85.30 31.24 91.41 **4)** 90.01 40.05 73.30 37.14 **5)** 80.45 41.42 44.60 75.50 **6)** 78.21 46.69 71.04 62.70
22	**1)** 45.43 60.22 69.25 77.30 **2)** 106.43 84.60 86.41 60.01 **3)** 104.01 108.03 95.36 61.23 **4)** 101.20 111.31 96.21 94.20 **5)** 72.21 118.00 111.30 84.20 **6)** 111.22 70.41 151.11 92.20
23	**1)** 12.33 14.22 12.60 12.24 **2)** 12.22 11.25 12.33 10.66 **3)** 13.74 12.02 13.94 16.02 **4)** 13.40 14.11 11.23 11.62 **5)** 11.26 11.53 13.24 16.06 **6)** 10.26 15.13 15.22 11.92
24	**1)** 14.23 12.00 11.64 13.02 **2)** 12.51 11.56 13.33 16.20 **3)** 15.34 14.33 15.43 11.68 **4)** 12.82 12.35 12.35 13.22 **5)** 14.30 9.62 11.16 11.90 **6)** 16.22 10.25 12.24 11.15
25	**1)** 97.06 97.17 109.26 112.11 **2)** 100.88 89.04 100.06 109.60 **3)** 85.48 104.01 79.63 100.52 **4)** 106.56 70.46 139.24 96.20 **5)** 113.24 84.25 127.30 99.11 **6)** 124.32 127.02 106.14 143.02
26	**1)** 133.25 63.26 92.11 139.23 **2)** 112.26 104.46 132.10 107.17 **3)** 122.45 137.12 110.65 137.20 **4)** 116.20 124.29 122.15 131.34 **5)** 208.23 221.69 192.41 234.47 **6)** 230.40 197.22 222.43 176.11
27	**1)** 5.57 **2)** 16.18 **3)** 38.42 **4)** 15.34
28	**1)** 11.90 **2)** 13.00 **3)** 24.40 **4)** 6.90 **5)** 15.00 **6)** 11.85 **7)** 19.80 **8)** 16.50 **9)** 23.00 **10)** 20.50 **11)** 19.50 **12)** 23.80 **13)** 16.95 **14)** 23.25
29	**1)** 4.85 9.70 **2)** 4.80 9.60 **3)** 5.25 15.75 **4)** 3.95 15.80 **5)** 2.30 11.50 **6)** 1.25 5.00 **7)** 3.80 19.00 **8)** 4.95 14.85 **9)** 8.25 33.00 **10)** 3.75 11.25 **11)** 1.18 5.90 **12)** 5.50 22.00 **13)** 4.25 12.75 **14)** 4.55 9.10
30	**1)** 12.10 **2)** 9.70 **3)** 11.95 **4)** 5.60 **5)** 9.30 **6)** 6.85 **7)** 16.65 **8)** 10.69
31	**1)** 5.95 **2)** 7.85 **3)** 7.00 **4)** 9.40 **5)** 5.79 **6)** 9.50 **7)** 7.40 **8)** 12.55

PG

32 **1)** 14.26 **2)** 6.89 **3)** 12.25 **4)** 14.53 **5)** 13.60 **6)** 14.30

33 **1)** 8.65 **2)** 6.95 **3)** 8.99 **4)** 8.55 **5)** 10.28 **6)** 49.92

34 **1)** 7.55 **2)** 8.49 **3)** 9.80 **4)** 9.40 **5)** 6.10 **6)** 13.55 **7)** 8.93

35 **1)** 7.38 **2)** 10.45 **3)** 10.20 **4)** 7.70 **5)** 7.80 **6)** 13.50

36 **1)** 6.39 **2)** 5.94 **3)** 6.09 **4)** 1.80 **5)** 6.07 **6)** 30.79

37 **1)** 19.35 **2)** 11.55 **3)** 18.84 **4)** 25.45 **5)** 18.15 **6)** 13.65

38 **1)** 37.75 **2)** 118 **3)** 312 **4)** 32 blocks **5)** 2,573.32 **6)** 110

39 **1)** 888.17 **2)** 191.75 **3)** 840.23 **4)** 502.57 **5)** 1,560.27 **6)** 2,075.70

40 *No answers necessary.*

41 *No answers necessary.*

42 **1)** .63 9.62 **2)** .67 10.23 **3)** .73 11.10 **4)** .77 11.77 **5)** .84 12.89 **6)** 1.04 15.84 **7)** 1.06 16.15 **8)** 1.17 17.84 **9)** 1.28 19.53 **10)** 1.36 20.82 **11)** 1.60 24.43 **12)** 1.73 26.51 **13)** 1.79 27.29 **14)** 1.86 28.39

43 **Top:** 10.00 + .70 = 10.70 9.50 +.67 = 10.17
Bottom: 11.40 +.80 = 12.20 9.69 + .68 = 10.37

44 **Top:** 9.20 + .64 = 9.84 11.55 + .81 = 12.36
Bottom: 11.95 + .84 = 12.79 9.63 + .67 = 10.30

45 **Top:** 17.05 + 1.19 = 18.24 16.50 + 1.16 = 17.66
Bottom: 19.14 + 1.34 = 20.48 10.70 + .75 = 11.45

46 **Top:** 8.20 + 1.23 + 1.25 + 4.50 = 15.18 + 1.06 = 16.24 5.80 + 7.25 + 1.80 + 2.30 + 2.85 = 20.00 + 1.40 = 21.40
Bottom: 7.90 + 1.04 + 1.04 + 1.15 + .90 = 12.03 +.84 = 12.87 7.60 + 4.25 + 3.95 + 3.75 = 19.55 + 1.37 = 20.92

47 **Top:** 1.35 + 1.60 + 6.00 + 2.20 = 11.15 + .78 = 11.93 2.25 + 2.30 + 4.60 + .90 + 1.15 = 11.20 + .78 = 11.98
Bottom: 16.50 + 2.30 + 2.95 + 3.10 = 24.85 + 1.74 = 26.59 8.00 + 4.80 + 1.18 + 1.10 + 1.04 = 16.12 + 1.13 =17.25

48 **Top:** 8.50 + 4.40 + 1.25 +.1.23 + 1.65 = 17.03 + 1.19 = 18.22 12.20 + 2.20 + 3.00 + 2.35 = 19.75 + 1.38 = 21.13
Bottom: 11.30 + 1.50 + 1.25 + 1.15 + .90 = 16.10 + 1.13 = 17.23 13.65 + 2.50 + 1.23 + 2.30 = 19.68 + 1.38=21.06

49 **Top:** 11.90 + 3.00 + 1.04 + 2.10 = 18.04 + 1.26 = 19.30 5.75 + 11.90 + 5.00 + 1.50 = 24.15 + 1.69 = 25.84
Bottom: 14.50 + 2.50 + 1.10 + .90 + 2.30 = 21.30 + 1.49 = 22.79 8.20 + 6.00 + 4.50 + 6.00 + 1.23 = 25.93 + 1.82 = 27.75

PG

50 **Top:** 11.00 + 6.50 + 1.15 + .90 + 2.20 = 21.75 + 1.52 = 23.27 14.55 + 3.75 + 2.35 + 2.85 + 2.30 = 25.80 +1.81 = 27.61
Bottom: 17.85 + 3.85 + 2.90 + 1.65 = 26.25 + 1.84 = 28.09 11.30 + 9.90 + 1.80 + 1.15 + 2.35 = 26.50 + 1.86 = 28.36

51 *Answers will vary.*

52 **1)** .04 .25 12.03
2) .74 .98 .69 3.56 8.77
3) 7.77 8.98 10.00 .71 .92
4) 1.11 1.22 1.31 8.61 8.73
5) 8.31 11.51 10.34 15.33
6) 15.58 70.11 81.51 105.23

53 **1)** 7.20 **2)** 18.23 **3)** 14.85 16.00 9.30 18.30 **4)** 615.79 **5)** 537.14

54 **1)** .24 .20 .24 .70 **2)** .17 .21 .64 .30
3) .31 .11 .11 .61 **4)** .12 .72 .30 .12
5) .86 .11 .20 .37 **6)** .44 .64 .44 .23

55 **1)** .11 .75 .34 .42 **2)** .31 .11 .17 .22
3) .51 .21 .42 .24 **4)** .11 .11 .16 .51
5) .31 .53 .71 .22 **6)** .53 .51 .22 .13

56 **1)** 1.67 7.11 1.14 1.28 **2)** 5.34 3.41 2.34 2.45
3) 1.10 5.70 6.31 1.14 **4)** 5.04 1.34 2.43 4.24
5) 5.11 3.10 2.22 2.22 **6)** 3.31 4.32 2.21 3.21

57 **1)** 4.32 2.40 2.04 7.16 **2)** 3.34 7.41 7.20 3.11
3) 1.34 5.43 2.41 6.66 **4)** 5.15 3.33 2.44 5.01
5) 4.02 2.34 1.45 1.11 **6)** 2.22 3.22 5.43 3.46

58 **1)** 1.11 2.04 2.11 12.01
2) 10.03 1.12 11.02 3.32
3) 12.75 10.43 2.03 4.14
4) 3.62 11.75 2.35 11.43
5) 3.52 12.21 21.12 31.56
6) 22.11 21.61 2.73 17.11

59 **1)** 83.25 40.24 13.17 31.48
2) 74.04 17.35 72.52 23.02
3) 12.14 62.04 56.67 10.23
4) 41.10 11.61 25.06 10.30
5) 21.26 10.41 53.45 75.25
6) 13.13 20.34 22.13 12.47

60 **1)** .17 .05 .58 .17 **2)** .09 .26 .16 .15
3) .18 .17 .39 .17 **4)** .66 .35 .28 .24
5) .45 .26 .36 .58 **6)** .09 .39 .29 .29

61 **1)** .59 .09 .29 .09 **2)** .17 .09 .09 .59
3) .57 .09 .19 .19 **4)** .19 .09 .09 .59
5) .57 .19 .38 .09 **6)** .19 .19 .13 .39

62 **1)** 1.85 2.79 5.78 4.89 **2)** .89 1.89 4.86 5.86
3) 1.87 2.88 2.87 4.78 **4)** .22 1.89 1.89 1.65
5) 1.44 2.89 5.89 4.69 **6)** .65 1.89 4.89 2.75

63 **1)** 2.79 3.85 .89 1.89 **2)** .17 3.77 3.69 5.45
3) 4.87 3.99 .57 1.16 **4)** 1.89 1.89 2.89 4.87
5) 2.89 5.89 .23 2.89 **6)** 1.45 3.77 1.59 5.09

64 **1)** 25.89 6.89 18.86 27.95
2) 8.95 27.85 27.85 8.99
3) 13.65 8.45 8.88 19.95
4) 36.98 39.09 58.58 20.96
5) 6.83 28.95 46.56 18.36
6) 8.89 18.88 47.36 9.88

Answer Key

PG	
65	**1)** 21.74 48.86 27.88 16.89 **2)** 5.96 18.89 18.89 68.57 **3)** 47.77 7.79 18.88 36.88 **4)** 68.88 28.89 18.79 27.69 **5)** 18.79 47.87 28.89 38.75 **6)** 47.89 27.38 46.33 34.56
66	**1)** 3.08 .11 6.11 3.44 **2)** 4.73 2.57 7.25 2.02 **3)** 3.11 3.61 5.09 .53 **4)** 5.31 4.27 5.16 1.03 **5)** 4.42 7.74 1.01 1.78 **6)** 1.01 2.05 2.34 2.21
67	**1)** 46.54 27.21 67.45 4.45 **2)** 3.25 33.41 66.32 53.02 **3)** 78.13 27.71 35.35 27.49 **4)** 7.14 36.22 44.13 56.38 **5)** 65.87 48.65 18.12 37.89 **6)** 53.44 74.09 25.53 3.12
68	**1)** 15.72 36.38 42.15 75.04 **2)** 1.64 7.85 7.13 24.11 **3)** 15.14 6.83 13.57 22.18 **4)** 6.69 3.48 8.11 16.19 **5)** 18.29 1.19 17.02 34.25 **6)** 17.86 17.78 46.31 8.88
69	**1)** 15.38 8.67 34.16 13.83 **2)** 38.84 27.81 53.15 4.58 **3)** 28.17 3.73 13.71 27.19 **4)** 25.08 5.17 12.26 43.74 **5)** 27.75 28.73 30.01 12.17 **6)** 52.11 43.68 4.13 13.17
70	**1)** .04 **2)** .75 **3)** 1.15 **4)** .05 **5)** 1.37 **6)** 3.91 **7)** .64 **8)** 4.65 **9)** 2.41 **10)** 1.25 **11)** 1.20 **12)** 1.64 **13)** 2.71 **14)** 3.93
71	**1)** .25 **2)** .05 **3)** .15 **4)** .80 **5)** 2.30 **6)** .75 **7)** 1.75 **8)** 2.15 **9)** 1.10 **10)** .80 **11)** 1.70 **12)** .15 **13)** 1.45
72	**1)** French Fries .10 **2)** Cola .02 **3)** Jumbo Burger 1.10 **4)** Vegetable Soup .10 **5)** Chili and Beans 1.60 **6)** Bacon and Eggs .50 **7)** Lemon Pie .85 **8)** Club Sandwich 1.10 **9)** Liver and Onions .45 **10)** Avocado Burger 1.10 **11)** Waffles .85 **12)** Clam Chowder .10 **13)** Milk Shake 1.96 **14)** Cheesecake .70
73	**1)** 1.10 **2)** 2.45 **3)** .70 **4)** .40 **5)** .25 **6)** 1.05 **7)** Jumbo Burger 1.30 **8)** 1.45
74	**1)** 1.85 **2)** .50 **3)** .75 **4)** 1.15 **5)** .15 **6)** .80 **7)** 2.20 **8)** 1.75
75	**1)** 6.59 **2)** 12 **3)** 265 **4)** 1,991 **5)** 1,550 **6)** 21.01
76	**1)** 164.79 **2)** 383.50 **3)** 1,731.00 **4)** Jennifer 68.25 **5)** 1,134.45 **6)** 725.35
77	**1)** 8.50 1.50 **2)** 3.75 1.25 **3)** 12.00 3.00 **4)** 16.50 3.50 **5)** 36.25 3.75 **6)** 19.50 .50 **7)** 19.20 30.80 **8)** 7.38 .62 **9)** 11.30 3.70 **10)** 29.75 10.25 **11)** 18.30 1.70 **12)** 23.80 26.20 **13)** 34.50 65.50 **14)** 16.50 33.50
78	**1)** 4.50 **2)** 3.41 **3)** 14.40 **4)** 7.35 2.65 **5)** 1.04 **6)** 8.63 11.37
79	**1)** 8.55 **2)** 9.25 **3)** 2.20 **4)** 2.16 **5)** .29 **6)** 1.85
80	**1)** 1.80 **2)** .82 **3)** 8.42 **4)** 43.25 **5)** 1.05 **6)** 1.45
81	**1)** 7.30 **2)** 2.10 **3)** 3.37 **4)** 25.05 **5)** 13.77

PG	
82	**1)** 8.50 + 4.50 + 2.50 = 15.50 (Change: 4.50) **2)** 5.95 + 5.75 + 2.30 = 14.00 (Change: 1.00) **3)** 16.50 + 6.20 + 2.70 = 25.40 (Change: 4.60) **4)** 9.10 + 3.69 + 6.00 + 6.10 + 2.30 = 27.19 (Change: 22.81)
83	**Top:** 14.00 + .98 = 14.98 (5.02) 13.00 + .91 = 13.91 (1.09) **Bottom:** 9.25 + .65 = 9.90 (.10) 10.50 + .74 = 11.24 (.76)
84	**Top:** 17.75 + 1.24 = 18.99 (1.01) 14.85 + 1.04 = 15.89 (4.11) **Bottom:** 10.43 + .73 = 11.16 (3.84) 10.65 + .75 = 11.40 (8.60)
85	**Top:** 16.35 + 1.14 = 17.49 (32.51) 18.86 + 1.32 = 20.18 (4.82) **Bottom:** 9.35 + .65 = 10.00 (10.00) 13.24 + .93 = 14.17 (5.83)
86	**Top:** 16.83 + 1.18 = 18.01 (1.99) 16.90 + 1.18 = 18.08 (.92) **Bottom:** 13.30 + .93 = 14.23 (.77) 19.75 + 1.38 = 21.13 (28.87)
87	**Top:** 7.25 + 5.95 + 1.10 + 1.15 + 3.30 = 18.75 + 1.31 = 20.06 (4.94) 4.10 + 4.55 + 5.25 + 2.70 = 16.60 + 1.16 = 17.76 (2.24) **Bottom:** 3.75 + 3.00 + 1.18 + 2.08 = 10.01 + .70 = 10.71 (4.29) 9.30 + 5.75 + 2.50 + 2.25 + 2.30 = 22.10 + 1.55 = 23.65 (6.35)
88	**Top:** 3.80 + 5.80 + 8.25 + 3.45 + 4.70 =26.00 + 1.82 = 27.82 (2.18) 5.80 + 13.00 + 3.80 + 3.60 = 26.20 + 1.83 = 28.03 (21.97) **Bottom:** 6.00 + 13.50 + 2.50 + 3.45 = 25.45 + 1.78 = 27.23 (2.77) 8.50 + 4.65 + 4.55 + 3.30 + 1.18 = 22.18 + 1.55 = 23.73 (16.27)
89	**Top:** 4.25 + 11.50 + 1.98 + 3.45 + 4.95 = 26.13 + 1.83 = 27.96 (22.04) 6.50 + 11.30 + 3.80 + 3.69 + 1.10 = 26.39 + 1.85 = 28.24 (1.76) **Bottom:** 11.50 + 8.50 + 6.15 = 26.15 + 1.83 = 27.98 (12.02) 9.60 + 9.00 + 4.16 + 1.04 = 23.80 + 1.67 = 25.47 (74.53)
90	**Top:** 9.10 + 11.90 + 2.40 + 3.30 = 26.70 + 1.87 = 28.57 (1.43) 9.00 + 9.90 + 4.16 + 2.70 = 25.76 + 1.80 = 27.56 (12.44) **Bottom:** 12.75 + 2.50 + 9.00 = 24.25 + 1.70 = 25.95 (14.05) 12.75 + 5.10 + 1.98 + 6.15 = 25.98 + 1.82 = 27.80 (72.20)
91	**1)** .74 .77 4.67 8.97 **2)** .90 6.74 9.21 15.36 **3)** 15.86 84.93 92.57 .33 **4)** 1.50 11.26 .29 3.87 **5)** 6.89 2.03 31.24 13.57 **6)** 31.09 57.63 39.89 3.12 **7)** 23.80 19.50 **8)** 14.35 **9)** 7.70 **10)** 36.00 **11)** 5.45 **12)** 681.60